Marc-Alexander Aßmann

Photon Statistics of Semiconductor Light Sources

Marc-Alexander Aßmann

Photon Statistics of Semiconductor Light Sources

Coherence properties of vertical-cavital surface-emitting lasers and polariton condensates

Südwestdeutscher Verlag für Hochschulschriften

Impressum/Imprint (nur für Deutschland/only for Germany)
Bibliografische Information der Deutschen Nationalbibliothek: Die Deutsche Nationalbibliothek verzeichnet diese Publikation in der Deutschen Nationalbibliografie; detaillierte bibliografische Daten sind im Internet über http://dnb.d-nb.de abrufbar.
Alle in diesem Buch genannten Marken und Produktnamen unterliegen warenzeichen-, marken- oder patentrechtlichem Schutz bzw. sind Warenzeichen oder eingetragene Warenzeichen der jeweiligen Inhaber. Die Wiedergabe von Marken, Produktnamen, Gebrauchsnamen, Handelsnamen, Warenbezeichnungen u.s.w. in diesem Werk berechtigt auch ohne besondere Kennzeichnung nicht zu der Annahme, dass solche Namen im Sinne der Warenzeichen- und Markenschutzgesetzgebung als frei zu betrachten wären und daher von jedermann benutzt werden dürften.

Coverbild: www.ingimage.com

Verlag: Südwestdeutscher Verlag für Hochschulschriften GmbH & Co. KG
Dudweiler Landstr. 99, 66123 Saarbrücken, Deutschland
Telefon +49 681 37 20 271-1, Telefax +49 681 37 20 271-0
Email: info@svh-verlag.de

Approved by: Dortmund, TU, Diss., 2010

Herstellung in Deutschland:
Schaltungsdienst Lange o.H.G., Berlin
Books on Demand GmbH, Norderstedt
Reha GmbH, Saarbrücken
Amazon Distribution GmbH, Leipzig
ISBN: 978-3-8381-2822-1

Imprint (only for USA, GB)
Bibliographic information published by the Deutsche Nationalbibliothek: The Deutsche Nationalbibliothek lists this publication in the Deutsche Nationalbibliografie; detailed bibliographic data are available in the Internet at http://dnb.d-nb.de.
Any brand names and product names mentioned in this book are subject to trademark, brand or patent protection and are trademarks or registered trademarks of their respective holders. The use of brand names, product names, common names, trade names, product descriptions etc. even without a particular marking in this works is in no way to be construed to mean that such names may be regarded as unrestricted in respect of trademark and brand protection legislation and could thus be used by anyone.

Cover image: www.ingimage.com

Publisher: Südwestdeutscher Verlag für Hochschulschriften GmbH & Co. KG
Dudweiler Landstr. 99, 66123 Saarbrücken, Germany
Phone +49 681 37 20 271-1, Fax +49 681 37 20 271 0
Email: info@svh-verlag.de

Printed in the U.S.A.
Printed in the U.K. by (see last page)
ISBN: 978-3-8381-2822-1

Copyright © 2011 by the author and Südwestdeutscher Verlag für Hochschulschriften GmbH & Co. KG and licensors
All rights reserved. Saarbrücken 2011

Contents

Motivation	**3**
1 Theoretical Background	**7**
1.1 Low-dimensional Semiconductor Structures	7
1.1.1 Carrier Eigenstates under reduced Dimensionality	7
1.1.2 Light-Matter Coupling in Semiconductor Nanostructures	9
1.2 Semiconductor Microcavities	12
1.2.1 Planar Microcavities	12
1.2.2 Micropillars	14
1.3 Quantum mechanical description of strong coupling	15
1.4 Classification of Light Fields	24
1.4.1 Coherent States	26
1.4.2 Thermal States	28
1.4.3 Fock States	30
2 Experimental Methods	**33**
2.1 Time resolved Correlation Spectroscopy	33
2.2 Optical Setup	36
2.2.1 The correlation streak camera technique	43
2.2.2 Characterization of Streak camera performance: time-integrated measurements	49
2.2.3 Characterization of Streak camera performance: time-resolved measurements	57
3 Quantum Dot VCSELs	**61**
3.1 QD VCSEL Samples	63
3.2 Correlation Measurements on QD VCSELs	65

3.3 Time-resolved Correlation Measurements 68

4 Quantum Well Diodes and VCSELs   81

4.1 Correlation measurements on QW VCSELs 83

5 Polaritonic Condensates   87

5.1 Nonequilibrium condensation 87

5.2 Strategies to reach degeneracy 87

    5.2.1 Definitions and signatures of BEC 89

5.3 Correlation measurements on polariton BECs and hip states 94

5.4 Dispersion measurements on polariton BECs and hip states 99

5.5 Nonresonantly pumped polariton BECs 102

6 Summary and Outlook   117

A Theoretical Model of QD Micropillar Lasers   119

Bibliography   125

List of figures   132

Index   135

Symbols and abbreviations   136

Acknowledgements   141

Motivation

My complete answer to the late 19th century question "what is electrodynamics trying to tell us?" would simply be this: Fields in empty space have physical reality; the medium that supports them does not.

Having thus removed the mystery from electrodynamics, let me immediately do the same for quantum mechanics: Correlations have physical reality; that which they correlate, does not.[1]

N. David Mermin

Modern telecommunications technologies rely heavily on optical data transmission. In accordance with the steadily growing demand for broadband internet bandwidths for everyday usage like high definition internet streaming and emerging services like cloud computing, there has been and will be a rapid increase in network traffic. Keeping pace with this expected rise in data transmission volumes in the following years will require faster optical networks able to cope with larger amounts of data. State-of-the-art devices allow data transfer rates as fast as 10 Gbps, but advancements are fast-paced. In these premises the Institute of Electrical and Electronic Engineers already promotes 100 Gigabit Ethernet as a standard for Ethernet that can transmit at speeds of 100 Gbps. Accordingly, there is a need for laser sources with low power consumption and fast modulation rates, which are stable under changes of external conditions like variations in temperature. Additionally, efficient interfaces of these optoelectronic devices to common electronics are heavily sought after, requiring a semiconductor-based approach. Especially attractive are solutions based on systems with reduced dimensionality. This reduction is reflected in a severe modification of the carrier density of states. As the free propagation of carriers is suppressed by introducing confinement in one or more dimensions, the density of carrier states changes accordingly, becoming discrete in the ultimate limit of zero-dimensional structures where no free carrier motion occurs at all. These structures showing full confinement are termed quantum dots and sometimes referred to as artificial atoms. However, in contrast to real atoms, their optical properties can be tailored on demand by changing the material system, quantum dot size and other parameters, which makes them very promising candidates for applications. Their discrete density of states makes them especially attractive for building efficient cavity-based light sources as any state not coupling to the cavity mode of interest represents a loss channel. Therefore, usage of quantum dots combined with cavity resonators with low mode volume allows one to build efficient light sources approaching the so-called thresholdless laser, which does not have any loss channels corresponding to spontaneous emission into other modes than the cavity mode of interest.

This thesis characterizes several industrially practicable solutions close to thresholdless lasing and compares their properties in terms of an appropriate experimental technique: ultrafast intensity correlation measurements. General theoretical considerations show that a proper characterization of such very efficient lasers requires complex experimental techniques to perform quantum optical analysis of photon statistics [2]. In chapter 1 a brief theoretical introduction into the basic principles of light-matter coupling in semiconductor nanostructures is given and the idea how to distinguish several possible states of a light field in terms of their photon statistics and correlations between photon emission events is introduced. Unfortunately, applying this techniques to semiconductor lasers requires detectors offering a temporal resolution on the order of picoseconds and either good efficiency or a very low dark count rate at the same time. Avalanche photo diodes, which are the commonly chosen detectors to perform these experimental techniques for atom-based lasers, where the need for high temporal resolution is lifted, can only fulfill one of these requirements at the same time. A different experimental approach to this problem is presented and discussed in detail in chapter 2. The basic idea of measuring photon statistics using a streak camera instead of avalanche photo diodes is introduced and benefits and drawbacks of using this method are discussed. Special emphasis is put on the effects of experimental imperfections like dark count rates, detector dead times and timing jitter issues. The following chapters deal with the application of the streak camera approach to several promising semiconductor light emitter concepts. Chapter 3 focuses on the coherence properties of quantum dot ensembles coupled to high-quality micropillar cavities, for which recently data transfer rates as high as 25 Gbps have been achieved [3]. These lasers offer stable output characteristics over a wide range of temperatures, thus removing the necessity to adjust the driving current to account for temperature-induced changes of the emission. This improvement results in more compact and efficient laser designs compared to earlier ones. For comparison, coherence properties of planar quantum well lasers are examined in chapter 4. Though in these structures carriers are only confined in one dimension and quantum well lasers are therefore less efficient than quantum dot lasers, they still find widespread application. Further advantages of using quantum well structures are designs allowing to tune the emission wavelength and offering large emission intensities. A refined approach to create efficient light sources based on quantum wells is described and investigated in chapter 5. When the carriers in quantum wells are strongly coupled to the photons inside the cavity mode, mixed quasiparticles of light and matter called polaritons become the eigenmodes of the coupled carrier-cavity system. The bosonic nature of these particles allows to achieve a large population of these quasiparticles in the ground state of the system without the need to achieve inversion conditions necessary in common semiconductor lasers. This process shares some similarities with Bose-Einstein condensation well known from diluted atomic gases and is therefore sometimes called a polariton Bose-Einstein condensate. However, there are also strong differences. Cavity polaritons are subject to dissipation and decay and therefore real equilibrium as realized in atomic condensates cannot be reached. Instead polaritons need to be inserted into the system constantly to replace the polaritons leaking from the cavity by their photonic content. This non-equilibrium nature of the condensed state gives rise to very rich physics and makes polaritonic condensates one of the most heavily investigated systems of the last few years and offers various possibilities to tailor the properties of the system in terms of the relative photonic content or spatial extent of the condensed state. Nevertheless, coherence properties of this system and its excitation spectrum are still not well known and understood. Therefore, systematic studies of these properties, which are in fact the quantities defining whether polariton condensates can really be considered on equal footing with their atomic counterparts, are presented and compared to several theoretical predictions.

Chapter 1

Theoretical Background

The efficient usage of semiconductors structures as light sources requires detailed control of the emission dynamics. As the name suggests, there are three main strategies to control light-matter coupling in semiconductors: Modifications of the photonic density of states, modifications of the carrier density of states and modifications of the coupling strength between them.

1.1 Low-dimensional Semiconductor Structures

This chapter summarizes basic theoretical concepts of tailoring the carrier density of states in terms of low-dimensional semiconductor structures and explains their optical properties.

1.1.1 Carrier Eigenstates under reduced Dimensionality

An analytical description of electronic motion inside a crystal must take Coulomb interactions with all electrons and all lattice ions into account. As the number N of carriers in a crystal is on the order of 10^{23} in macroscopic structures, the corresponding set of coupled one-particle Schroedinger equations

$$\hat{H}\psi_i(\vec{r}_i) = E_i\psi_i(\vec{r}_i) \qquad (i = 1\ldots N) \tag{1.1}$$

with

$$\hat{H} = \sum_{i=1}^{N} \hat{H}_{kin}^{(i)} + \sum_{i=1}^{N}\sum_{j=1}^{N_{ion}} \hat{H}_{e^{(i)}-ion^{(j)}} + \sum_{i,k=1;i\neq k}^{N} \hat{H}_{e^{(i)}-e^{(k)}} \tag{1.2}$$

represents a problem unsolvable by computational means unless further assumptions are made. It is possible to decouple those equations by treating the collective effects of carriers and nuclei in terms of a mean average crystal potential $U_{mean}(\vec{r})$. This leads to a set of Schroedinger equations using an individual carrier single-particle Hamiltonian:

$$\{-\frac{\hbar^2}{2m}\nabla^2 + U_{mean(\vec{r})}\}\psi(\vec{r}) = E\psi(\vec{r}). \quad (1.3)$$

Considering the periodicity of the crystal lattice allows for further simplification. The mean average crystal potential will share the periodicity of the crystal. Therefore

$$U_{mean(\vec{r})} = U_{mean(\vec{r}+\vec{a})} \quad (1.4)$$

where $\vec{a}$ is a basic translation between lattice sites. The Eigenstates will then take the form of Bloch waves

$$\psi_{\vec{k}}(\vec{r}) = u_{\vec{k}}(\vec{r})e^{i\vec{k}\vec{r}} \quad (1.5)$$

which are composed of a periodic envelope function and a plane wave. The finite crystal size leads to a discretization of $\vec{k}$ in all dimensions:

$$k_i = \frac{2\pi n_i}{a_i N_i}; \qquad -\frac{N_i}{2} < n_i \leq +\frac{N_i}{2} \quad (i = x, y, z), \quad (1.6)$$

where n_i is an integer number. For bulk crystals the large number of elemental lattice cells creates a quasi-continuum of allowed Bloch wavevectors.

The effective-mass approximation allows further simplification by treating the carriers as quasi-free particles. The effective mass of electrons and holes in most general form is an anisotropic tensor with components given by the shape of the conduction and valence bands, respectively:

$$\frac{1}{m^*_{e,h;l,m}} = \frac{1}{\hbar^2}\frac{\partial E_{c,v}(\vec{k})}{\partial k_l \partial k_m} \qquad (l, m = x, y, z). \quad (1.7)$$

For carriers near the center of the first Brillouin zone the dispersions for both conduction and valence bands approach an isotropic and parabolic shape. In the parabolic band approximation the effective masses of electrons and holes therefore become constant numbers, yielding carrier eigenenergies in formal analogy to the free carrier case:

$$E_{e,h}(\vec{k}) = \frac{\hbar^2 |k|^2}{2m^*_{e,h}}. \quad (1.8)$$

The carrier eigenstates change drastically if their motion is constrained in one or more dimensions on the order of the De Broglie wavelength

$$\lambda^{dB}_{e,h} = \frac{2\pi\hbar}{\sqrt{3m^*_{e,h}k_B T}}, \quad (1.9)$$

where k_B denotes the Boltzmann constant and T the temperature. In this case quantization effects become prominent, which manifest e.g. in the density of states giving the number of available carrier eigenstates within an interval dE around energy E. For electrons in the conduction band this dependency is given by (see Fig.1.1):

$$D_{3D}(E) = \frac{1}{2\pi^2}\left(\frac{2m_e^*}{\hbar^2}\right)^{\frac{3}{2}}\sqrt{E - E_{CB}} \qquad (1.10)$$

$$D_{2D}(E) = \frac{m_e^*}{\pi\hbar^2}\sum_i \Theta(E - E_{CB} - \Delta E_i) \qquad (1.11)$$

$$D_{1D}(E) = \frac{1}{\pi}\sqrt{\frac{2m_e^*}{\hbar^2}}\sum_i \left(\frac{n_i \Theta(E - E_{CB} - \Delta E_i)}{\sqrt{E - E_{CB} - \Delta E_i}}\right) \qquad (1.12)$$

$$D_{0D}(E) = \sum_i n_i \delta(E - E_{CB} - \Delta E_i), \qquad (1.13)$$

where ΔE_i gives the difference between the conduction band minimum and the i-th energy level in the conduction band and n_i gives the degeneracy of that energy level. It is apparent that increasing carrier confinement results in quantization of the energy levels in the confined direction.

Up to now the properties of individual carriers have been considered. In optical experiments the excitation of an electron from the valence to the conduction band also creates a hole in the valence band. However, due to the strong confinement in low-dimensional heterostructures like quantum wells (2D), quantum wires (1D) and quantum dots (0D) the Coulomb interaction between them becomes important, resulting in bound electron-hole pair states called excitons. The binding energy of this quasiparticle depends on the material and the dimensionality of the semiconductor structure, but is usually in the meV or eV range.

1.1.2 Light-Matter Coupling in Semiconductor Nanostructures

The electron and hole constituting an exciton form a dipole which interacts with the electromagnetic light field. In the simplifying picture of one photon mode interacting with one excitonic mode in a two-coupled-oscillator model [4] the complex frequency eigenvalues $\omega_{U,L}$ of the coupled system are expressed as follows:

$$\omega_{U,L} = \frac{\omega_x + \omega_c}{2} - i\frac{\gamma_x + \gamma_c}{2} \pm \sqrt{M^2 + \frac{1}{4}(\omega_x - \omega_c - i(\gamma_x - \gamma_c))^2}, \qquad (1.14)$$

where ω_x and ω_c are the bare exciton and photon frequencies and γ_x and γ_c are the exciton and photon mode damping rates, respectively. M is the corresponding coupling matrix element of the two oscillators. In the case of vanishing coupling $M \to 0$ the eigenfrequencies of the exciton and photon modes are recovered as expected. Specializing to the resonant case $\omega_x = \omega_c$, however, two different kinds of behaviour can be seen for nonvanishing coupling

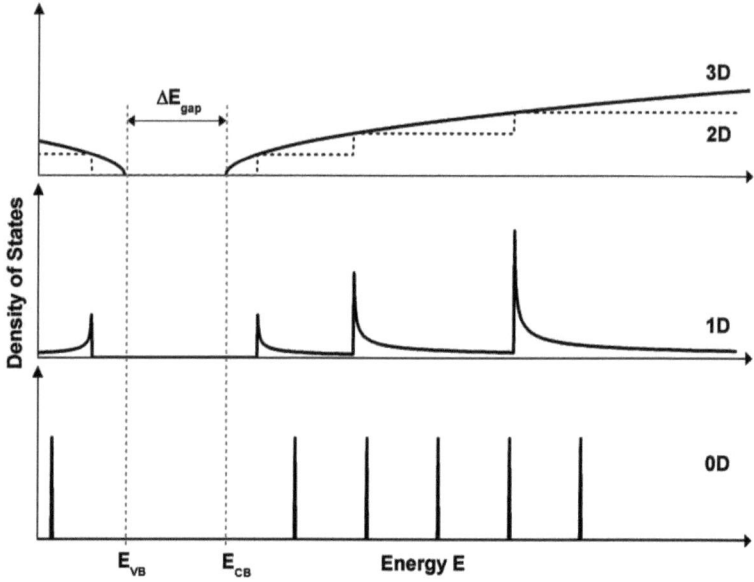

Figure 1.1: Comparison of the semiconductor individual carrier density of states for different dimensionalities. The reduction of the degrees of freedom manifests in a discretization of the states.

of the oscillator modes. In the so-called weak coupling regime $2M < |\gamma_x - \gamma_c|$ the square root is completely imaginary. Accordingly the real parts of the coupled oscillator solution still give the eigenfrequencies of the bare exciton and photon modes, but the decay rates are altered significantly, resulting in an enhancement of the exciton decay rate. In the strong coupling regime $2M > |\gamma_x - \gamma_c|$ the square root becomes entirely real and a Rabi splitting $\Omega_R = 2\sqrt{|M|^2 - \frac{1}{4}(\gamma_c - \gamma_x)^2}$ between the eigenfrequencies of the coupled system occurs. The crossover from weak to strong coupling is illustrated in Fig.1.2 as a function of the dipole interaction matrix element. The splitting occurs in the imaginary part of the frequency in the weak coupling and in the real part of the frequency in the strong coupling regime. Obviously both the splitting in the strong coupling regime and the enhanced decay rate in the weak coupling regime are of interest for applications in photonics and for manifacturing tailored light sources. However, in order to reach these regimes the strength of the interaction between exciton and

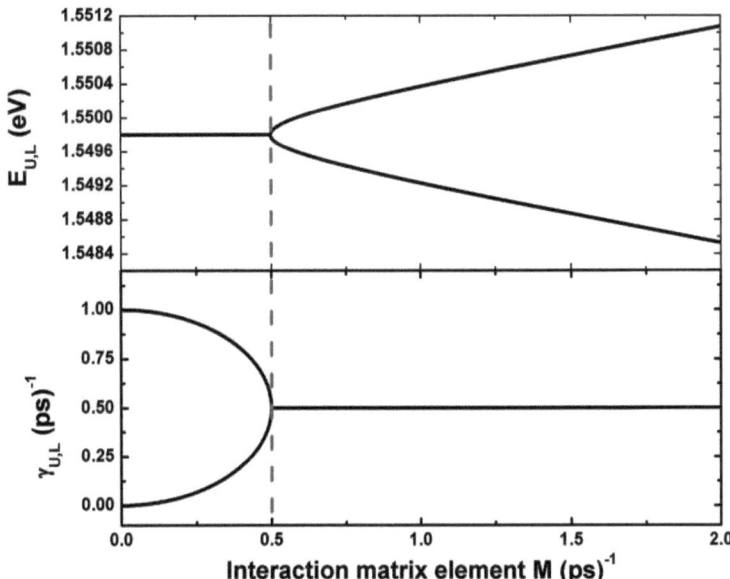

Figure 1.2: Transition from weak to strong coupling with increasing interaction matrix element. The red dashed line indicates the boundary between both regimes. The upper panel shows the splitting of the eigenmodes in the strong coupling regime. The lower panel shows the enhanced decay rate of the excitonic mode in the weak coupling regime.

photon modes needs to be increased significantly compared to the bulk case. Semiconductor nanostructures offer two possibilities to address this issue: Confining the light field by introducing a microcavity, which will be discussed in the next chapter, and increasing the exciton oscillator strength by decreasing the exciton Bohr radius.

Modifications of the exciton Bohr radius rely on the interaction matrix element M dependence on the overlap integral of the electron and hole envelope functions χ integrated over the whole volume [5]:

$$M \propto \int_V d\vec{r}\, \chi^*_{c,k_e}(\vec{r}) \chi_{v,k_h}(\vec{r}). \tag{1.15}$$

For free electron and holes without any confinement potential the envelope functions are plane waves,

$$\chi_{c,k_e} = \exp(i\vec{k_e}\vec{r}) \quad (1.16)$$

$$\chi_{v,k_h} = \exp(i\vec{k_h}\vec{r}). \quad (1.17)$$

In this case the overlap integral reduces to the momentum conservation condition $\delta(\vec{k_e}-\vec{k_h})$. For an exciton the envelop function consists of a plane wave for the center of mass motion with momentum K and the bound relative motion. Here the momentum conservation condition becomes $\vec{K} = 0$ and the overlap integral is enhanced by $\sqrt{\frac{V^3}{a_b}}$ [6] as long as the volume of interest allows coherent addition of the dipoles which is usually the case if the size of the volume is comparable to the wavelength of the em field interacting resonantly with the exciton. Unlike free electrons and holes moving independently through the volume V, the electron and hole move together with an average separation a_b depending on the confinement, increasing the probability of an optical transition. Still, this relation also opens up another problem. The total oscillator strength also depends on the total volume. Especially for quantum dots with sizes in the range of 100 nm, microcavitites are additionally used to increase the light-matter coupling.

1.2 Semiconductor Microcavities

1.2.1 Planar Microcavities

While the possibility of changing the electronic density of states has already been discussed, another strategy to modify the light-matter interaction lies in tailoring the photonic density of states. A convenient way to realize this lies in using an optical resonator. The resonant optical modes will be confined while other modes will be suppressed inside the cavity. The frequency width of the resonant enhancement is characterized by the quality factor

$$Q = \frac{\omega_m}{\delta\omega_m} \quad (1.18)$$

which is the ratio of a resonant cavity mode frequency to the linewidth of the mode. Equivalently, the lifetime of a photon in a cavity is given by $\tau_m = \frac{Q}{\omega_m}$. One of the main problems of enhancing the exciton-photon interaction rate in nanostructures is the small volume of the nanostructure, in which the interaction can take place. To increase the efficiency of the coupling it is necessary to keep the ratio of the interaction volume to the total volume of the cavity as large as possible. This is achieved by using microcavities: cavities with a length close to the dimension of the wavelength of light. The small volume of the semiconductor nanostructure leads to a need for extremely high reflectivities beyond 99 % in order to achieve efficient exciton-photon coupling [7]. Metallic mirrors are in most cases not sufficient to reach such high reflectivities. A common choice to realize better confinement of the photons are distributed Bragg reflectors. These structures consist of many pairs of multilayers of materials with different refraction index surrounding a central cavity. For most of the structures examined in this thesis the chosen materials and refractive indices will be GaAs ($n_{GaAs} = 2.95$) and AlAs ($n_{AlAs} = 3.65$). These materials have almost identical lattice constants allowing growth of heterostructures by means of molecular beam epitaxy (MBE). The thickness of each of the layers is chosen such that the optical path length in each of the layers is a quarter of the designed central wavelength λ of the cavity mode and each of the double layers shows a reflectivity determined by $\frac{n_{GaAs}}{n_{AlAs}} \approx 0.808$. The central cavity is usually chosen as a λ-cavity with a length of $L_R = \frac{\lambda}{n_{GaAs}}$. The electromagnetic field inside the cavity can be described as a standing wave with antinodes at the center of the cavity and at the interfaces of the cavity to the DBR-structures. Accordingly the em-field will penetrate into the mirrors [8], resulting in an effective penetration length of

$$L_{DBR} = \frac{\lambda}{2n_{GaAs}} \frac{n_{GaAs} n_{AlAs}}{n_{AlAs} - n_{GaAs}} \tag{1.19}$$

and a total effective cavity length of $L_{eff} = L_R + L_{DBR}$ [9]. The total reflectivity of the structures is given by a broad spectral region of high reflectivity centered around λ which is called stop band. In the stop band the total reflectivity depends on the number N of pairs of mirror layers as [10]:

$$R = 1 - 4\frac{1}{n_{GaAs}} \left(\frac{n_{GaAs}}{n_{AlAs}}\right)^{2N}. \tag{1.20}$$

Under imperfect growth conditions the center frequency of the stop band ω_s and the frequency corresponding to the length of the cavity ω_c are not necessarily identical. In this case the cavity mode frequency is given by [11]:

$$\omega_m = \frac{L_R \omega_c + L_{DBR} \omega_s}{L_{eff}}. \tag{1.21}$$

As the penetration depth into the mirrors is usually larger than the cavity length this means that the central mode frequency is mostly determined by the center of the stop band and not by the cavity length.

The planar microcavities described up to now rely on axial confinement causing a fixed axial wavevector $k_z = \frac{2\pi}{L_R}$. There is no confinement perpendicular to the growth axis, so the cavity photon will have an in-plane dispersion leading to a cavity photon energy of approximately

$$E_0 = \frac{\hbar c}{n_{GaAs}}\sqrt{k_z^2 + k_\parallel^2} \qquad (1.22)$$

For small $k_\parallel$ this is a parabolic dispersion and can be described by the cavity photon gaining some effective mass

$$m_{cav} = \frac{h n_{GaAs}}{cL_R}. \qquad (1.23)$$

The resulting effective cavity photon mass is very small. Usually it is about five orders of magnitude smaller than the electron mass [12]. It should be noted that the calculations above work very well for empty cavities. More complicated structures containing emitters are handled in terms of transfer matrix theory.

1.2.2 Micropillars

Further reduction of the mode volume can be achieved by confining the em field inside a cavity in all three dimensions. This is realized by combining electron beam lithography with plasma etching to create cylindrical pillars of varying diameter from a planar microcavity. In the simplifying case of perfectly reflecting sidewalls the cavity modes are then given by: [13]

$$E_{n_x,n_y} = \sqrt{E_0^2 + \frac{\hbar^2 c^2}{n_{GaAs}^2}(k_{x,n_x}^2 + k_{y,n_y}^2)}, \qquad (1.24)$$

where the lateral wavenumbers are given by $k_{i,n_i} = \pi(n_i+1)/L_{Lat}$, L_{Lat} is the lateral size of the cavity, $i = x,y$ and n_i are integer quantum numbers starting from 0. For the most common case of a cylindrical pillar the mode spectrum is given by:

$$E = \sqrt{E_0^2 + \frac{\hbar^2 c^2}{n_{GaAs}^2}\frac{\eta_{n_\varphi,n_r}^2}{R_{cav}^2}}. \qquad (1.25)$$

Here R defines the radius of the pillar and η_{n_φ,n_r} is the n_rth zero of the Bessel function $J_{n_\varphi}(\eta_{n_r,n_\varphi}r/R)$. Obviously the mode energies become blueshifted with decreasing pillar diameter due to stronger confinement. It becomes possible to tune the fundamental mode energy by choosing an appropriate pillar diameter during growth of the structure. Further, the free spectral range between two non-degenerate modes of a micropillar which is of importance for the efficiency of micropillar lasers, increases with decreased diameter. However, also the losses due to scattering at the micropillar sidewalls and intrinsic losses increase for small pillar diameters [14], resulting in a reduced Q factor. Therefore it is not feasible to reduce the pillar diameter far below 1 μm. Taking applications into account it should also be noted that the fundamental mode of a circular micropillar is twofold polarization degenerate.

1.3 Quantum mechanical description of strong coupling

Although strong coupling has also been reported for quantum dots in a microcavity [15] this section will focus on quantum wells in planar microcavities. A naive classical picture of the strong coupling regime of two oscillators has been given in section 1.1.2. Nevertheless, a thorough description of strong light-matter coupling requires a fully quantum-mechanical picture. This section will follow the approach given in [16]. Neglecting off-resonant terms the Hamiltonian for a coupled system of light and excitons becomes

$$H = \hbar\omega_C \hat{a}^\dagger \hat{a} + \hbar\omega_X \hat{b}^\dagger \hat{b} + \hbar g(\hat{a}\hat{b}^\dagger + \hat{a}^\dagger \hat{b}). \tag{1.26}$$

The detuning of the eigenenergies is given by

$$\hbar\Delta = \hbar(\omega_C - \omega_X). \tag{1.27}$$

Mode $\hat{a}$ describes cavity photons as introduced in 1.2.1 and is a pure Bose operator which obeys the commutation relation

$$[\hat{a}\hat{a}^\dagger] = 1. \tag{1.28}$$

Mode $\hat{b}$ describes the excitation in the material and can be anything from a pure Bose to a pure Fermi operator depending on the kind of excitation considered. For excitons $\hat{b}$ becomes approximately bosonic in the low-density limit and the Hamiltonian 1.26 represents two linear oscillators coupled with interaction strength g. This problem can be described in the basis $|i,j\rangle_{bare}$ with i being the number of excitations in the material and j being the number of photons. These states are called bare states and this description gives an intuitive picture of the excitation and loss processes as in a microcavity usually the losses are bare photons escaping through the DBR mirrors and the matter excitations are directly created by external pumping. In the Heisenberg picture the time dependence is given by

$$\begin{bmatrix} \hat{a}(t) \\ \hat{b}(t) \end{bmatrix} = \exp(-i\omega t) \begin{bmatrix} \cos\frac{Gt}{2} - i\frac{\Delta}{G}\sin\frac{Gt}{2} & -2i\frac{g}{G}\sin\frac{Gt}{2} \\ -2i\frac{g}{G}\sin\frac{Gt}{2} & \cos\frac{Gt}{2} + i\frac{\delta}{G}\sin\frac{Gt}{2} \end{bmatrix} \begin{bmatrix} \hat{a}(0) \\ \hat{b}(0) \end{bmatrix}, \tag{1.29}$$

where $\omega = (\omega_X + \omega_C)/2$ and $G = \sqrt{4g^2 + \Delta^2}$. Obtaining observables is now straightforward. The population for a state with initial condition $|\psi\rangle = \kappa|1,0\rangle_{bare} + \zeta|0,1\rangle_{bare}$ is given by

$$(\hat{a}^\dagger \hat{a})(t) = |\kappa|^2 \cos^2\left(\frac{Gt}{2}\right) + \frac{|\Delta\kappa + 2\zeta g|^2}{G^2}\sin^2\left(\frac{Gt}{2}\right) - \Im(\kappa\zeta^*)\frac{4g}{G}\cos\left(\frac{Gt}{2}\right)\sin\left(\frac{Gt}{2}\right). \tag{1.30}$$

The oscillatory terms in the population and the off-diagonal terms in 1.29 already show that a system prepared in a bare state will not stay in this state, but evolve. The evolution is plotted

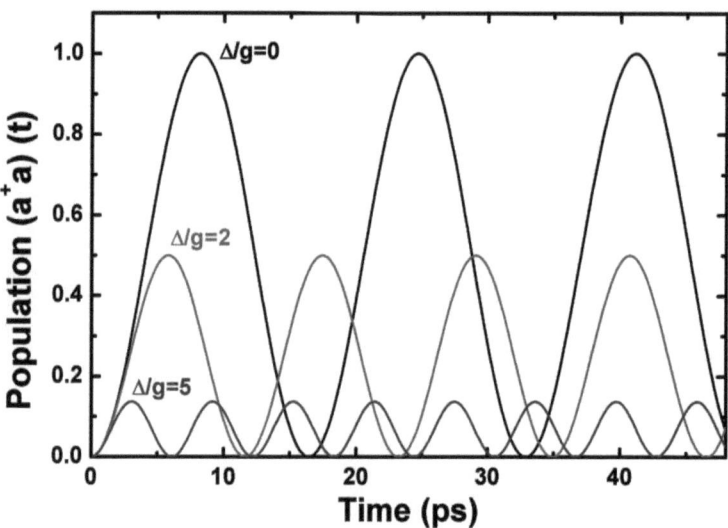

Figure 1.3: Population $(\hat{a}^\dagger \hat{a})(t)$ for a state prepared with initial condition $|\psi_0\rangle = |1,0\rangle_{bare}$ as a function of time for different detunings. The population undergoes Rabi oscillations. Detuning increases the transition rate, but lowers the conversion efficiency.

in figure 1.3 for a state with $\zeta = 0$ and $\kappa = 1$. The population oscillates back and forth between the bare states at the generalized Rabi frequency G, resulting in the so-called Rabi oscillations. Another approach is to diagonalize the Hamiltonian 1.26 in a basis of dressed states. In the Schrödinger picture the most general substitution is given by

$$\hat{p} = \alpha \hat{a} + \beta \hat{b} \quad (1.31a)$$
$$\hat{q} = \gamma \hat{a} + \delta \hat{b}. \quad (1.31b)$$

$\alpha, \beta, \gamma, \delta$ are complex quantities. $\hat{p}$ and $\hat{q}$ shall remain bosonic operators, therefore

$$[\hat{p}, \hat{p}^\dagger] = [\hat{q}, \hat{q}^\dagger] = |\alpha|^2 + |\beta|^2 = |\gamma|^2 + |\delta|^2 = 1 \quad (1.32)$$

and

$$[\hat{p}, \hat{q}] = [\hat{p}, \hat{q}^\dagger] = \alpha \gamma^* + \beta \delta^* = 0 \quad (1.33)$$

The bare states can be expressed in this new basis as

$$\hat{a} = \frac{\delta \hat{p} - \beta \hat{q}}{\alpha \delta - \beta \gamma} \tag{1.34a}$$

$$\hat{b} = \frac{-\gamma \hat{p} + \alpha \hat{q}}{\alpha \delta - \beta \gamma}. \tag{1.34b}$$

Substituting those into 1.26 yields

$$(\alpha\delta - \beta\gamma)^2 \hat{H} = \hat{p}^\dagger \hat{p} \ [\hbar\omega(|\delta|^2 + |\gamma|^2) - \hbar g \Im(\delta^*\gamma)] \tag{1.35a}$$

$$+ \ \hat{q}^\dagger \hat{q} \ [\hbar\omega(|\beta|^2 + |\alpha|^2) - \hbar g \Im(\beta^*\alpha)] \tag{1.35b}$$

$$+ \ \hat{p}^\dagger \hat{q} \ [\hbar\omega(-\beta\delta^* - \gamma^*\alpha) + \hbar g(\alpha\delta^* + \beta\gamma^*)] + h.c. \tag{1.35c}$$

Diagonalizing the Hamiltonian requires the terms in line 1.35c to be zero. This condition is fulfilled if

$$\alpha\delta^* + \beta\gamma^* = 0. \tag{1.36}$$

All the requirements are met when choosing $\alpha = \cos(\theta)$, $\beta = \sin(\theta)$, $\gamma = -\sin(\theta)$ and $\delta = \cos(\theta)$. The final form of $\hat{p}$ and $\hat{q}$ is given by

$$\hat{p} = \cos(\theta)\hat{a} + \sin(\theta)\hat{b} \tag{1.37a}$$

$$\hat{q} = -\sin(\theta)\hat{a} + \cos(\theta)\hat{b}, \tag{1.37b}$$

where

$$\cos(\theta) = \frac{\Delta + G}{\sqrt{2\Delta^2 + 8g^2 + 2\delta G}}. \tag{1.38}$$

θ is the so-called mixing angle. The Hamiltonian 1.26 is now diagonalized:

$$\hat{H} = \hbar\omega_p \hat{p}^\dagger \hat{p} + \hbar\omega_q \hat{q}^\dagger \hat{q}. \tag{1.39}$$

The eigenfrequencies of the eigenmodes are given by

$$\omega_{p/q} = \omega \pm \frac{G}{2} \tag{1.40}$$

and plotted in figure 1.4. This diagonalization approach is basically what is known as Bogoliubov transformation.

In the following dressed states will be denoted as $|n, m\rangle$ with n dressed particles of energy $\hbar\omega_p$ and m particles of energy $\hbar\omega_q$. It is now straightforward to define sets of states with fixed total number of excitations N. These read

$$H_N = \{|n, m\rangle, n + m = N\}, \tag{1.41}$$

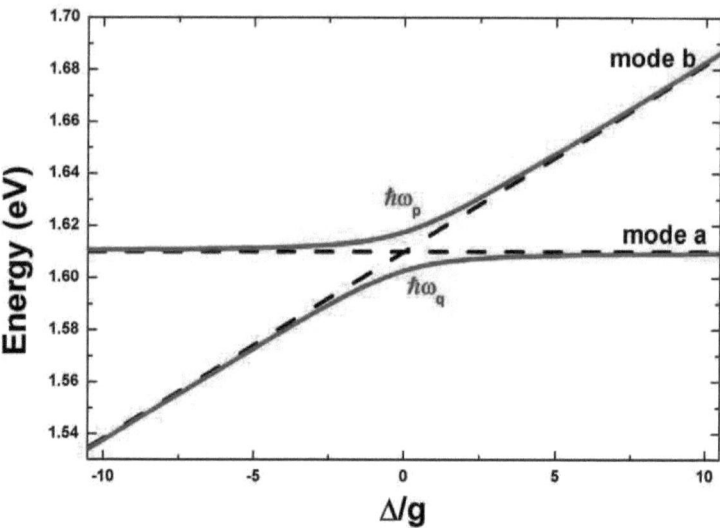

Figure 1.4: Eigenenergies (solid lines) of the system given by 1.26 as a function of the detuning δ given in multiples of the interaction strength g. At zero detuning an anticrossing is observed. For large detunings the bare modes (dashed lines) are recovered.

where n and m are non-negative integers. Any excitation escaping the system by photon leakage out of the cavity or nonradiative exciton recombination can be seen as a transition from H_N to H_{N-1}. These processes remove one excitation from one of the oscillators and only transitions from $|n,m\rangle$ to $|n-1,m\rangle$ or $|n,m-1\rangle$ are allowed, carrying away energy amounts of $\hbar\omega_p$ and $\hbar\omega_q$, respectively. It is a bit surprising that the emission of a photon carries away the energy of a dressed particle instead of the energy of a bare photon. This behavior shows that the quantum mechanical system is in full analogy to the classical case where the coupled system behaves like two independent oscillators with frequencies ω_p/ω_q. To clarify what this means it is instructive to examine the experimentally most relevant case of vacuum Rabi splitting and neglect the experimentally also observable higher rungs [17]. Here H_1 is coupled to the single vacuum mode, resulting in a doublet of transitions. One can calculate the amplitudes of these transitions in terms of the bare state annihilation operators as:

$$M_1 = \langle 0,0|\hat{a}|1,0\rangle = \alpha(\Delta/g) \tag{1.42a}$$

$$M_2 = \langle 0,0|\hat{a}|0,1\rangle = \gamma = -\alpha(-\Delta/g) \tag{1.42b}$$

$$M_3 = \langle 0,0|\hat{b}|1,0\rangle = \beta = \alpha(-\Delta/g) \tag{1.42c}$$

$$M_4 = \langle 0,0|\hat{b}|0,1\rangle = \delta = \alpha(\Delta/g). \tag{1.42d}$$

Those amplitudes are shown in figure 1.5. The physical quantitity of interest is the square of theses amplitudes:

$$|X|^2 = |\alpha(\Delta/g)|^2 \tag{1.43a}$$

$$|C|^2 = |\alpha(-\Delta/g)|^2. \tag{1.43b}$$

These are the so-called Hopfield coefficients [18]. They can be considered as the relative weight

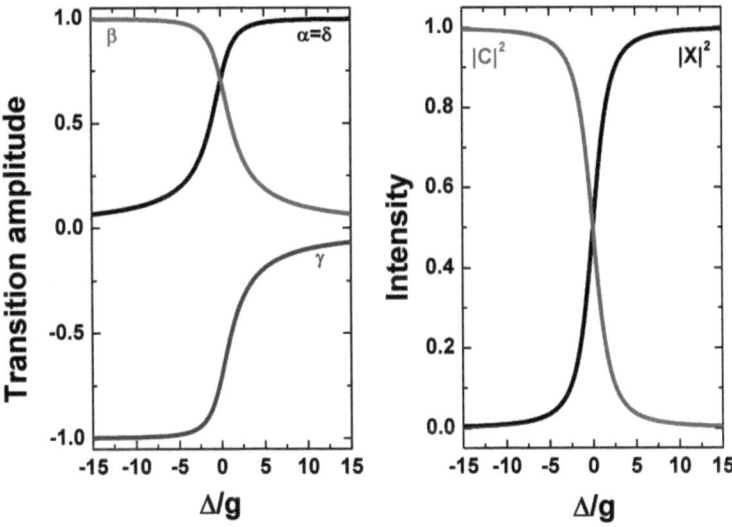

Figure 1.5: Transition amplitudes between the vacuum and dressed states with one excitation (left) and corresponding intensities (right) as a function of the detuning.

of the bare states in the dressed states. Accordingly one can interpret the dressed states as quasiparticles composed of excitons and photons with relative photonic and excitonic content depending on the detuning. In the quasiparticle picture the mode with energy $\hbar\omega_q$ is called lower polariton. Its excitonic fraction $|X|^2$ gets larger with increased detuning, while its photonic fraction $|C|^2$ gets smaller. Due to the antisymmetry of the transition amplitudes with respect to the detuning, the mode with energy $\hbar\omega_p$ behaves the other way around and is called upper polariton. Its excitonic fraction is given by $|C|^2$ and gets smaller with increased detuning, as shown in figure 1.5. Accordingly it is possible to precisely adjust the excitonic and photonic contents of polaritons by changing the detuning.

In order to fully describe microcavity polaritons, it is now necessary to take the microcavity dispersion into account. The energy of cavity photons depends on their in-plane wavevector $k_{||}$. Accordingly also the corresponding polariton energies will show a dispersion depending on $k_{||}$. This dispersion will depend on the energy mismatch

$$\Delta_k = E_C(k_{||}) - E_X(k_{||}) \tag{1.44}$$

between the bare cavity photon dispersion E_C (equation 1.22) and the bare exciton dispersion which is given by

$$E_X^n = E_{gap} + E_e(k_{||}) + E_h(k_{||}) + E_n. \tag{1.45}$$

Here E_{gap} is the band gap energy of the material, E_e and E_h are the electron and hole dispersions given by equation 1.8 and E_n is the binding energy of the bound state with quantum number n, treated in analogy to the hydrogen atom. The dispersions for the upper and lower polariton branches are then given by

$$E_{\substack{U\\L}}(k_{||}) = \frac{1}{2}(E_X(k_{||}) + E_C(k_{||})) \pm \frac{1}{2}\sqrt{\Delta_k^2 + 4\hbar^2 g^2(k_{||})}, \tag{1.46}$$

respectively. Also the Hopfield coefficients become dependent on $k_{||}$:

$$|C_k|^2 = \frac{1}{2}(1 - \frac{\Delta_k}{\sqrt{\Delta_k^2 + 4\hbar^2 g^2}}) \tag{1.47a}$$

$$|X_k|^2 = \frac{1}{2}(1 + \frac{\Delta_k}{\sqrt{\Delta_k^2 + 4\hbar^2 g^2}}) \tag{1.47b}$$

and therefore also the excitonic and photonic fractions of polaritons will depend on $k_{||}$. Accordingly also the lifetime $\tau_{LP/UP}$ of polaritons is composed of the excitonic and photonic lifetimes τ_X and τ_C and will show a detuning-dependent variation on $k_{||}$[19]

$$\frac{1}{\tau_{LP}(k_{||})} = \frac{|C_k|^2}{\tau_C} + \frac{|X_k|^2}{\tau_X}. \tag{1.48}$$

The lower polariton becomes most photonic and the lower polariton lifetime therefore becomes close to the cavity photon lifetime at small $k_{||}$, especially for negative detuning as shown in figure

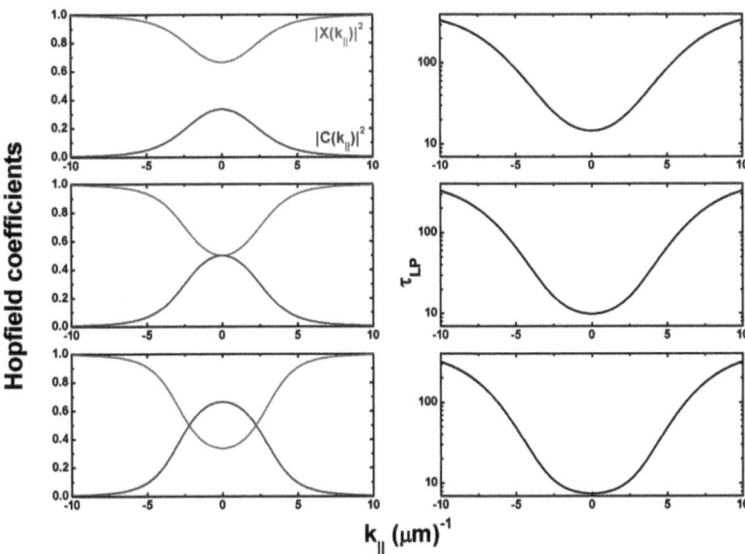

Figure 1.6: Left panel: Hopfield coefficients as a function of $k_{\parallel}$ for $2\hbar g = 14.4\,\text{meV}$. The coefficients are displayed for detunings of $+5\,\text{meV}$ (top), $0\,\text{meV}$ (middle) and $-5\,\text{meV}$ (bottom). Right panel: Corresponding lower polariton lifetimes for the same detunings assuming a cavity lifetime of $5\,ps$ and an exciton lifetime of $500\,\text{ps}$.

1.6. For large $k_{\parallel}$ the lower polariton becomes almost completely excitonic. For comparison equation 1.46 is plotted in figure 1.7 for positive, negative and no detuning. The exciton-photon interaction results in an avoided crossing of the polariton dispersion. In this simple treatment the appearance of this anticrossing is independent of the strength of the interaction. A more realistic model is given by taking also the broadening of exciton and photon resonances into account by adding imaginary components to the bare exciton and cavity photon dispersions 1.22 and 1.45 in a similar manner as in the classical treatment of equation 1.14. The broadening terms are given by $-i\gamma_X$ and $-i\gamma_C$, respectively. γ_X is the broadening induced by exciton interactions with phonons or other particles and γ_C reflects the finite linewidth caused by the finite reflectivity of the microcavity. This linewidth depends inversely proportional on the microcavity quality factor. For zero detuning the photon and exciton states are exchanging

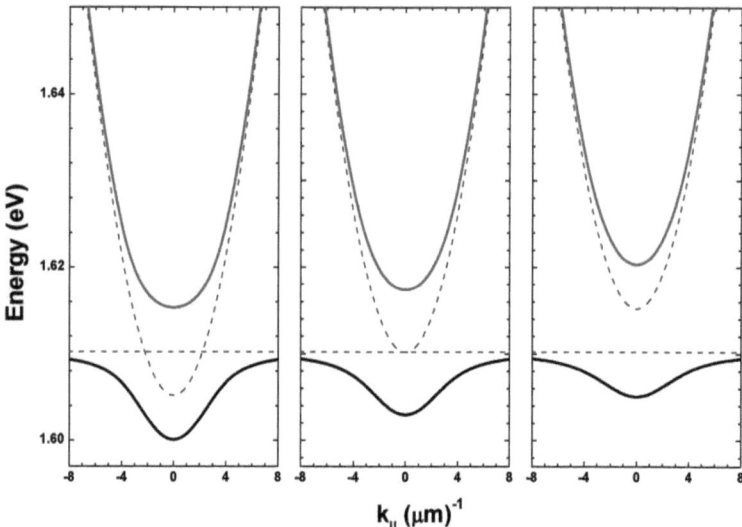

Figure 1.7: Exciton-polariton energies at detunings of −5 meV (left), 0 meV (middle) and +5 meV (right). Solid lines correspond to the lower and upper polariton dispersion, respectively. Dashed lines give the eigenmodes of the bare system.

energy at the Rabi frequency Ω_R and equation 1.46 then becomes

$$E_{\substack{U\\L}}(k_\parallel) = \frac{1}{2}(E_X(k_\parallel) + E_C(k_\parallel) - i\gamma_X - i\gamma_C) \pm \frac{1}{2}\sqrt{4\hbar^2\Omega_R^2 - (\gamma_X - \gamma_C)^2}. \quad (1.49)$$

Obviously this expression will depend strongly on the sign of the expression below the square root $\hbar\Omega_L = \sqrt{4\hbar^2\Omega_R^2 - (\gamma_X - \gamma_C)^2}$. If $2\hbar\Omega_R > |\gamma_X - \gamma_C|$, $E_{\substack{U\\L}}$ exhibits a Rabi splitting of the system eigenmodes in analogy to the classical treatment. This behavior is a sign of the strong coupling regime. It should be mentioned that this splitting is a purely theoretical quantity and does not necessarily coincide with the splittings measured in experiments. In particular, there will be different splittings seen in transmission, absorption, reflectivity and photoluminescence measurements [20]. In the high-reflectivity limit these are given by

$$\hbar\Omega_T = 2\sqrt{\sqrt{\hbar^4 G^4 + 2\hbar^2 G^2 \gamma_X(\gamma_X + \gamma_C)} - \gamma_X^2} \quad (1.50a)$$

$$\hbar\Omega_A = 2\sqrt{\hbar^2 G^2 - \frac{1}{2}(\gamma_X^2 + \gamma_C^2)} \quad (1.50b)$$

$$\hbar\Omega_R = 2\sqrt{\sqrt{\hbar^4 G^4 (1 + \frac{2\gamma_X}{\gamma_C})^2 + 2\hbar^2 G^2 \gamma_X^2 (1 + \frac{\gamma_X}{\gamma_C})} - 2\hbar^2 G^2 \frac{\gamma_X}{\gamma_C} - \gamma_X^2} \quad (1.50c)$$

$$\hbar\Omega_{PL} = \sqrt{2\hbar\Omega_L \sqrt{\hbar^2 \Omega_L^2 + (\gamma_X + \gamma_C)^2} - \hbar^2 \Omega_L^2 - (\gamma_X + \gamma_C)^2}. \quad (1.50d)$$

Those splittings and $\hbar\Omega_L$ are plotted in figure 1.8. Obviously none of the experimentally

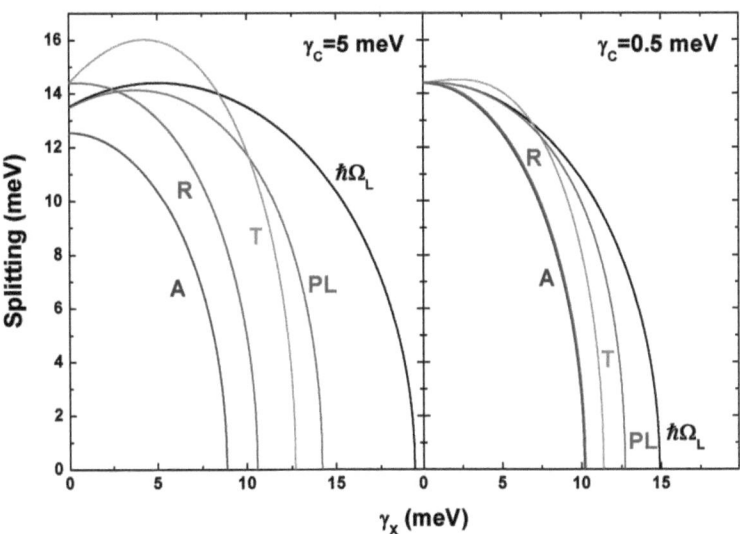

Figure 1.8: The splittings as seen in the five different quantities mentioned in the text for $2\hbar\Omega_R=14.4$ meV as a function of the exciton broadening. The left panel corresponds to a bad cavity with a cavity mode line width of 5 meV. The lower panel shows the results for a better cavity with a line width of 0.5 meV.

obtainable splittings corresponds to $\hbar\Omega_L$. It is therefore not trivial to show unambiguously that a system is indeed in the strong coupling regime as a splitting seen in the experiment is not necessarily proportional to $\hbar\Omega_L$. However, it can be shown that the following relation holds

[21]:

$$\hbar\Omega_T \geq \hbar\Omega_R \geq \hbar\Omega_A, \tag{1.51}$$

and also $\Omega_L \geq \Omega_A$. The splitting in absorption is always a sufficient condition to ensure that the system is in the strong coupling regime. In the good cavity limit the splitting in photoluminescence can be a good approximation of $\hbar\Omega_L$, but should not be taken as a proof of the strong coupling regime.

If $\hbar\Omega_R < |\gamma_X - \gamma_C|$ the square root becomes imaginary and the energy splitting disappears. Such behavior marks the weak-coupling regime where a description in terms of weakly interacting bare cavity photons and excitons is appropriate and the energies of the bare modes can become degenerate at some $k_\parallel$ as already seen in the classical description given in section 1.1.2.

1.4 Classification of Light Fields

Besides obvious parameters like frequency, polarization or intensity there are also further characteristics of light fields which manifest in their coherence properties, which can be described by a hierarchy of correlation functions stating with the field-field correlation function

$$g^{(1)}(\vec{r}_1, t_1, \vec{r}_2, t_2) = \frac{\langle E^-(\vec{r}_1, t_1) E^+(\vec{r}_2, t_2) \rangle}{\sqrt{\langle |E(\vec{r}_1, t_1)|^2 \rangle \langle |E(\vec{r}_2, t_2)|^2 \rangle}}. \tag{1.52}$$

E^- and E^+ denote the negative and positive frequency parts of a mode of the light field, respectively. $g^{(1)}$ is a measure of phase correlations of a light field and reflects in the contrast of interference patterns of the em field. The two common quantities deduced from $g^{(1)}$ are the coherence time τ_{coh} and the correlation length l_{coh} which give the time and distance over which phase correlations are maintained, respectively. Still, a complete characterization of em fields which is also able to identify nonclassical states requires consideration of correlation functions of at least second order. Neglecting any spatial dependencies the normal-ordered second-order photon number correlation function is given by

$$g^{(2)}(t_1, t_2) = \frac{\langle \hat{a}^\dagger(t_1) \hat{a}^\dagger(t_2) \hat{a}(t_1) \hat{a}(t_2) \rangle}{\langle \hat{a}^\dagger(t_1) \hat{a}(t_1) \rangle \langle \hat{a}^\dagger(t_2) \hat{a}(t_2) \rangle}, \tag{1.53}$$

where $\hat{a}^\dagger$ and $\hat{a}$ are the photon creation and destruction operators for the photon mode of interest, respectively. The normal-ordering assures that the change of the state of the em field introduced by the detection of a photon is taken into account. For stationary light sources $g^{(2)}$ will only depend on the relative delay τ between two photon detections:

$$g^{(2)}(\tau) = \frac{\langle \hat{a}^\dagger(t) \hat{a}^\dagger(t+\tau) \hat{a}(t) \hat{a}(t+\tau) \rangle}{\langle \hat{n}(t) \rangle \langle \hat{n}(t+\tau) \rangle}, \tag{1.54}$$

where the averages are time averages and $\hat{n}$ denotes the photon number of the mode of interest. $g^{(2)}$ can be considered as the conditional probability to detect a photon at a delay τ after the detection of a first photon, normalized to the probability of a second detection for photons which are emitted statistically independent of each other. For very large delays $\tau \to \infty$ the

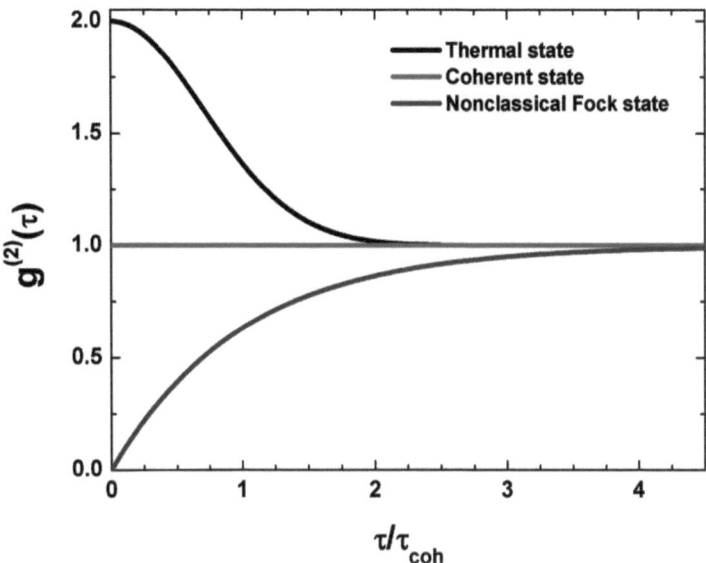

Figure 1.9: Second order correlation function $g^{(2)}(\tau)$ for thermal, coherent and nonclassical light. τ is measured in multiples of the coherence time τ_{coh} of the light field.

photon emission events are uncorrelated in any state of the light field, so $g^{(2)}(\tau \to \infty) = 1$. It is possible to distinguish three basic kinds of states of the light field, namely thermal, coherent and nonclassical light, by comparing $g^{(2)}(\tau = 0)$ to $g^{(2)}(\tau \to \infty)$ depending on whether the probability for simultaneous detection of two photons is increased, unaltered or decreased. The enhanced or decreased photon pair detection probability relaxes back towards 1 on a timescale depending on the coherence time of the light as shown in Fig.1.9. Accordingly the value of the equal time second-order correlation function

$$\begin{aligned}g^{(2)}(0) &= \frac{\langle n(t)(n(t)-1)\rangle}{\langle n(t)\rangle^2} \\ &= 1 - \frac{1}{\langle n(t)\rangle} + \frac{\langle(\Delta n(t))^2\rangle}{\langle n(t)\rangle^2}\end{aligned} \quad (1.55)$$

is a good characterization of the state of the light field. It is composed of three terms: The first term is a unity valued constant. The negative second term describes the change of the state of the light field induced by the detection of the first photon. The positive third term takes the intrinsic noise of the photon emission process into account in terms of the relative photon number variance. Correlation functions can be generalized up to arbitrary order to describe the probability of n-photon detections. The most general definition of a nth order correlation function is given by

$$g^{(n)}(t_1\ldots t_n) = \frac{\langle: \prod_{i=1}^{n} \hat{n}(t_i) :\rangle}{\prod_{i=1}^{n} \langle \hat{n}(t_i)\rangle}. \quad (1.56)$$

The double stops denote normal ordering of the underlying photon creation and annihilation operators. The statistical properties of coherent, thermal and nonclassical light will be discussed in more detail in the following sections.

1.4.1 Coherent States

In a classical picture both amplitude and phase of a wave are well defined. Any quantum approach to a description of the light field must, however, be in accordance with the Heisenberg time-energy uncertainty relation

$$\Delta E \Delta t \geq \hbar \quad (1.57)$$

or equivalently the uncertainty relation for photon number n and phase ϕ

$$\Delta n \Delta \phi \geq 1. \quad (1.58)$$

Coherent states are one realization of minimum uncertainty states ($\Delta n \Delta \phi = 1$) with the uncertainties distributed equally among the two quadratures

$$\hat{x} = \frac{\hat{a} + \hat{a}^\dagger}{\sqrt{2}} \quad (1.59)$$

$$\hat{p} = \frac{\hat{a} - \hat{a}^\dagger}{\sqrt{2}i} \quad (1.60)$$

of the field and the closest approach to a classical wave picture without any uncertainties. Another similarity of coherent states to the classical description is found in their immunity to loss. Considering only single modes, coherent states $|\alpha\rangle$ are eigenstates of the annihilation operator $\hat{a}$:

$$\hat{a}|\alpha\rangle = \alpha|\alpha\rangle \qquad (1.61)$$

with complex eigenvalues $\alpha = |\alpha|\exp(i\phi)$. The statistical properties of coherent states can be calculated by using the photon number states $|n\rangle$ as a basis and expressing coherent states as a superposition of photon number states

$$\begin{aligned}|\alpha\rangle &= \sum_n |n\rangle\langle n|\alpha\rangle \\ &= \sum_n \exp\left(-\frac{1}{2}|\alpha|^2\right)\frac{\alpha^n}{\sqrt{n!}}|n\rangle. \end{aligned} \qquad (1.62)$$

It is now possible to calculate the photon number probability distribution in this basis as

$$\begin{aligned}P_{coh}(n) &= |\langle n|\alpha\rangle|^2 \\ &= \left|\exp\left(-\frac{1}{2}|\alpha|^2\right)\sum_m \frac{\alpha^m}{\sqrt{m!}}\langle n|m\rangle\right|^2 \\ &= \exp\left(-|\alpha|^2\right)\frac{|\alpha|^{2n}}{n!}\end{aligned} \qquad (1.63)$$

which corresponds to a Poisson distribution with mean photon number $\langle n\rangle_{coh} = |\alpha|^2$ as shown in the upper panel of 1.11. For this distribution the most probable photon number coincides with the mean photon number. The Poisson distribution describes statistically independent events which is another characteristic similar to the classical description. This behavior also reflects in the equal time correlation function as defined in equation 1.55. The variance of a Poissonian process is given by $\langle(\Delta n_{coh})^2\rangle = \langle n\rangle$. Accordingly the third term in equation 1.55 becomes equal to $\frac{1}{\langle n\rangle}$ and exactly cancels out the effect of the second term describing the effect the detection of the first photon had on the light field. Only the constant factor

$$g^{(2)}_{coh}(0) = 1. \qquad (1.64)$$

is left, indicating that the photon emission and detection events are indeed statistically independent for coherent states regardless of the intensity of the mode. This result remains unaltered for correlations up to arbitrary order:

$$g^{(n)}_{coh}(0) = 1. \qquad (1.65)$$

Any state of the light field can be decomposed into a superposition of coherent states. However, coherent states are in general not orthogonal [22]. Therefore they form an overcomplete basis.

1.4.2 Thermal States

The concept of thermal or chaotic radiation refers to a radiation field which is in thermal equilibrium with a black body acting as an emitter. This means that emission and absorption of the black body cancel out. Accordingly the corresponding quantized radiation field can be considered to have an effective temperature T matching that of the black body. The spectrum of thermal radiation follows an universal form given by Planck's law with temperature acting as the only parameter. For thermal radiation the spectral energy density per unit volume is given by

$$u(\omega)d\omega = \frac{\hbar}{\pi^2 c^3} \frac{\omega^3 d\omega}{\exp(\hbar\omega/k_B T) - 1}. \quad (1.66)$$

where k_B is the Boltzmann constant. At a fixed temperature the photon number distribution for a single mode of the light field is given by a Bose-Einstein distribution of known mean photon number:

$$P_{th}(n) = \frac{\langle n \rangle^n}{(\langle n \rangle + 1)^{n+1}}. \quad (1.67)$$

This distribution is shown for several mean photon numbers in figure 1.11. Independent of the mean photon number the most probable photon number is always 0 which is a striking feature of the Bose-Einstein distribution. Accordingly thermal states of the light field show larger photon number fluctuations compared to coherent states which manifest in the variance and the higher order moments of the Boltzmann distribution. The variance is given by $\langle (\Delta n_{th})^2 \rangle = \langle n \rangle^2 + \langle n \rangle$. Inserting this result into equation 1.55 shows that similar to the coherent case all photon number dependent contributions cancel out and a constant result remains. However, due to increased photon number fluctuations the result is twice as high as in the coherent case:

$$g_{th}^{(2)}(0) = 2. \quad (1.68)$$

The resulting photon bunching is a signature of thermal states. In a classical pictures this effect is caused by strong fluctuations of the momentary photon number around the mean photon number. Detection of a photon means that it is very probable that the momentary intensity is much larger than the mean intensity. Accordingly the probability to detect another photon is increased. Calculation of higher order correlation functions can be performed by evaluating the higher order factorial moments of the Boltzmann distribution. It turns out that the simultaneous detection probability shows a factorial dependence on the number of photons simultaneously detected:

$$g_{th}^{(n)}(0) = n!. \quad (1.69)$$

Explaining this effect in a quantum mechanical picture is more difficult, but can be done in terms of interference of indistinguishable probability amplitudes [23]. This explanation is

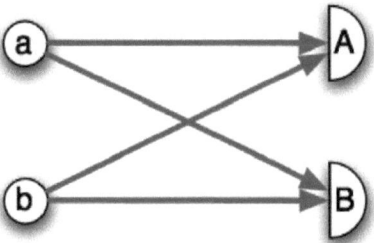

Figure 1.10: Schematic picture of interfering two-photon probability amplitudes.

illustrated in figure 1.10. There are two particle sources a and b and two detectors A and B. Assuming that each source can only emit one particle at a time, there are two possibilities to realize simultaneous detections at A and B. Either a particle is emitted at a and detected at A and another particle is emitted at b and detected at B (red arrows) or vice versa (green arrows). The corresponding probability amplitudes describing the processes of particle emission at one source and registration at one detector are termed $D_{a,A}$, $D_{b,B}$, $D_{a,B}$ and $D_{b,A}$, respectively. The probability of a simultaneous detection event in terms of these probability amplitudes is given by

$$P_{dist} = |D_{a,A} D_{b,B}|^2 + |D_{b,A} D_{a,B}|^2 \qquad (1.70)$$

if the two possible events leading to the simultaneous detection are distinguishable. If those events are indistinguishable the situation changes as the probability amplitudes must now be added causing a different probability of a simultaneous detection:

$$P_{indist} = |D_{a,A} D_{b,B} + D_{b,A} D_{a,B}|^2$$
$$= P_{dist} + D_{a,A} D_{b,A}^* D_{b,B} D_{a,B}^* + D_{a,A}^* D_{b,A} D_{b,B}^* D_{a,B}. \qquad (1.71)$$

The magnitude of the additional interference term is exactly as large as P_{dist} and can be brought into the same form by exchanging particles, but its sign depends on the symmetry of the wavefunction of the particle considered. For bosonic particles the wavefunction does not change under exchange of particles resulting in the interference term being equal to P_{dist}, while for fermionic particle the wavefunction will change sign and the interference term will exactly cancel out P_{dist}. As a result this quantum mechanical approach can explain bosonic bunching and fermionic antibunching behavior in the same framework. Further, the factorial dependence of photon bunching can be considered as a consequence of the number of possible permutations of indistinguishable probability amplitudes.

1.4.3 Fock States

Fock states are eigenstates of the photon number operator $\hat{n}$ and form a complete set of orthogonal states. The photon number distribution for a Fock state shows sub-Poissonian character: For a k-photon Fock state $P_{Fock}(n = l)$ is exactly 1 for $l = k$ and 0 for $l \neq k$ as depicted in figure 1.11. This reflects in a vanishing variance $(\Delta n_{Fock})^2 = 0$ and a photon number dependent

$$g^{(2)}_{Fock}(0) = 1 - \frac{1}{\langle n \rangle}. \tag{1.72}$$

For a single-photon Fock state $g^{(2)}_{Fock}(0)$ reduces to 0 indicating the quantized nature of light. In fact, any $g^{(2)}(0)$ smaller than one cannot be explained by a classical probability distribution and characterizes non-classical states of the light field. As a consequence of the uncertainty

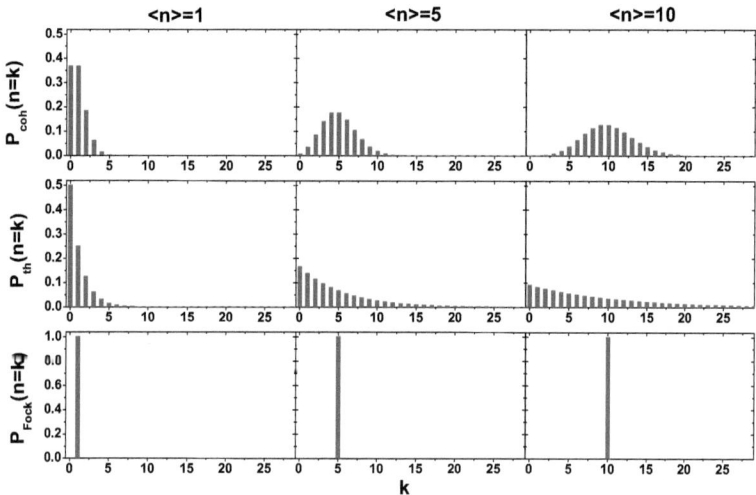

Figure 1.11: Photon number distributions for coherent, thermal and Fock states for different mean photon numbers.

principle photon number eigenstates must also reveal a completely undefined phase.

Chapter 2

Experimental Methods

Intensity correlation spectroscopy is a well known technique which has been used in different disciplines from astronomy to biology to perform diverse tasks like determination of the angular diameters of stars [24] and diffusion coefficients of particles undergoing Brownian motion [25], identification of soft matter surface fluctuations [26], optical coherence tomography [27] and classification of photon sources [28]. The main idea of this technique lies in gathering information about a photon number distribution without measuring the absolute photon number distribution. As no detector shows perfect efficiency, direct measurements of these distributions are almost impossible. For most practical purposes, it is sufficient to know relative characteristic quantities of the photon number distribution like the relative value of the detected photon number fluctuation to the mean photon number given by equation 1.55. These quantities are easier to access experimentally. Common techniques aiming at measuring these fluctuations rely on measuring the correlations between the intensities of two beams of light and their temporal and spatial variations. These are directly linked to the variance of the photon number distribution and allow to draw further conclusions on the underlying photon number distributions.

Below, several experimental realizations of this approach are presented and compared. Especially the conditions under which the different approaches give accurate results are discussed in detail.

2.1 Time resolved Correlation Spectroscopy

Time resolved correlation spectroscopy and time-correlated single photon counting (TCSPC) are techniques for measuring low-level signals at high repetition frequency with high temporal resolution. Although they can be applied to continuous wave signals, this section will focus on pulsed emission. The basic idea is to measure the distribution of single photon arrival times inside the signal period and build up a histogram of the photon arrival times relative to the beginning of the signal period. By doing so, the time resolution is given by the transit time spread in the detector and not by the width of the single-photon pulses. Usually single-photon avalanche photodiodes (SPADs) are used as detectors. They are subject to a deadtime after the detection of a photon during which no additional photon detections are possible. Accordingly, the number of photons detected per signal period should be very low, usually on the order of 0.01 to 0.1 per pulse. Otherwise the so-called pile-up effect occurs: The photon detection probability at late times inside the signal will be reduced by the frequent detections of photons at early times in the pulse which block the detector until the end of the pulse period. In its simplest form, this technique can be used to reconstruct a waveform with high temporal resolution. A schematic of this method is shown in figure 2.1. However, determination of photon statistics as discussed in section 1.4 is difficult using just a single SPAD because the crucial information are photon pairs. These can only be accessed if both the dead time and the SPAD time resolution are much shorter than the coherence time. The dead time limitation can be overcome by performing a measurement in Hanbury-Brown-Twiss (HBT) configuration [29]. Here the incoming beam is split into two beams and each is detected by a SPAD. The signals are fed to a time-to-amplitude converter, which creates a histogram of the delay times between two photon detection events. If pile-up effects are avoided, the number of coincidence counts for delays inside one period compared to the number of coincidence counts for delays, where the photons stem from different periods and are therefore uncorrelated, is a measure of $g^{(2)}(0)$. Still two problems remain: If the signal coherence time is shorter than the SPAD time resolution or if the presence of multiple photons at the detector is very probable, the measured $g^{(2)}(0)$ will be distorted. The HBT setup is most commonly used for characterization of single photon sources in terms of photon antibunching. Here these two drawbacks are not significant. As only one photon is expected to be present in each period, multiphoton processes and coherence time issues are not crucial. As long as the signal count rates are large compared to the background noise count rates, the HBT-approach will measure $g^{(2)}(0)$ correctly for single photon sources. For thermal light the situation is more difficult. Usually there are many photons present and for thermal emission from semiconductor structures the coherence time lies usually in the range of tens of picoseconds, while the temporal resolution of efficient SPADs is on the order of 400 ps. Additionally the photon number emitted per pulse is fluctuating strongly, so that even when the detected photon number per pulse is small, the pile-up effect can occur during the periods where large photon numbers are present. In fact, even for ideal detectors and no background

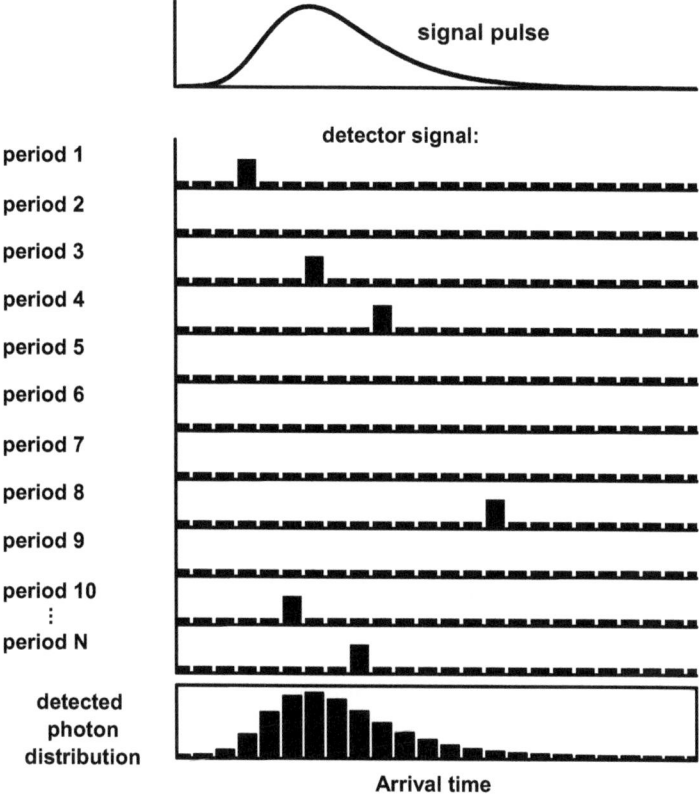

Figure 2.1: Principle of a time-correlated single photon counting experiment. Inside each pulse there is either no detection or a single detection of a photon. After N repeated detections of signal periods the distribution of arrival times follows the waveform of the signal for low detection probability.

noise, the inability of SPADs to distinguish between single photon detections and multiphoton detections causes the measured $g^{(2)}(0)$ to differ from the real one for thermal light [30]. The following sections will describe a different approach aimed at measuring correlation functions of semiconductor light sources especially in the thermal regime and the optical setup used to realize it.

2.2 Optical Setup

The optical setup used for the streak camera intensity correlation measurements and the momentum-space dispersion measurements is shown in Fig. 2.2. The optical excitation of the studied samples is provided by laser pulses with a duration of either 1.5 ps or approximately 100 fs at a pulse repetition rate of 75.39 MHz. These pulses were created with a mode-locked Ti:sapphire laser (Coherent Mira 900-D, see figure 2.3) pumped by a 10 W Nd:YVO$_4$ CW pump laser (Coherent Verdi V-10) emitting at 532 nm. The mode-locked laser emission is tunable in a wavelength range from 700 to 980 nm. Optionally, a frequency doubling unit based on second harmonic generation can be used to realize pulses in the wavelength range of 400 - 500 nm. The mode-locked laser can be configured to emit pulses with a duration of approximately 2 ps or 100 fs. The excitation beam optics set is shown in figure 2.4. A small fraction of the laser emission is detected by a fast photo diode to create a trigger signal for the streak camera. Well-defined control of the excitation power density is achieved by using variable neutral density attenuators. A Glan-Taylor-prism is used to polarize the beam. The extinction ratio of the prism is on the order of 10^5. A $\lambda/2$ and $\lambda/4$ waveplate allow to create any linear, circular or elliptical polarization. The excitation laser light is focused onto the sample surface either using a lens with a focal length of $f = 65$ mm under a freely chosen incidence angle. Alternatively, excitation under normal incidence is possible using a 92 %/8 % beam splitter and a microscope objective. In this case, the transmitted 92 % of the excitation power are dumped, while the 8 % reflected intensity are directed towards the sample. This configuration allows for very small spot diameters on the sample. Microscope objectives as shown in figure 2.5 with different magnification rates and numerical apertures of 0.26 (10x objective), 0.42 (50x objective) and 0.5 (100x objective) are used. The position of the microscope objective can be precisely adjusted by piezo-based actuators. Spot sizes on the sample are on the order of few μm for normal incidence excitation and about 20 μm using the lens. The samples are mounted on a cold finger in a helium-flow cryostat depicted in figure 2.6. The temperature can be set in a range of $T = 6 - 300$ K by controlling the helium-flow rate and using external heating. The Cryostat and

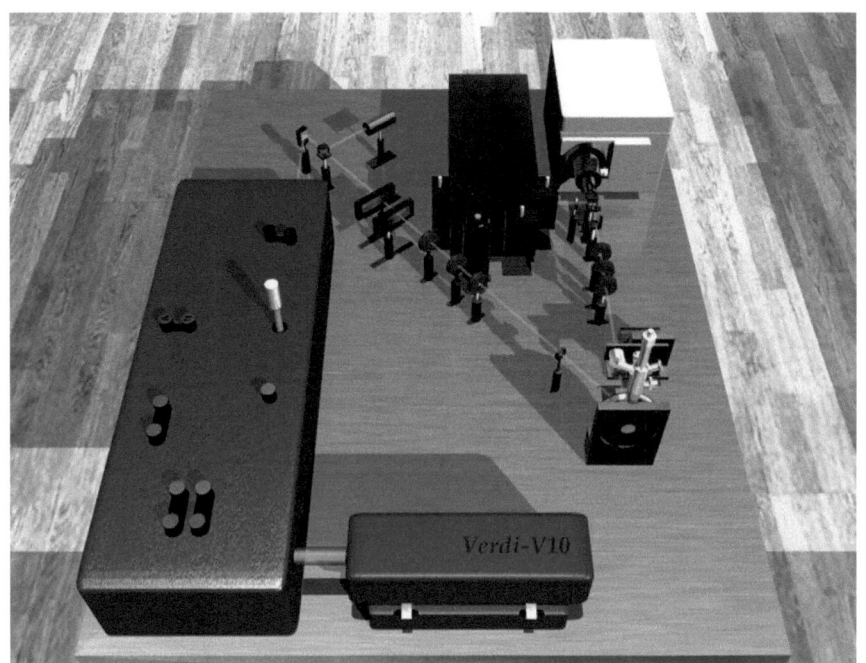

Figure 2.2: Overview of the experimental setup used for the streak camera intensity correlation and momentum-space dispersion measurements. A detailed description of the single components grouped into several functional units is given in the text below and in the figure captions of figures 2.3 to 2.9. The functional units discussed are the laser system consisting of the CW pump laser and the tunable Ti:sapphire laser, the optics in the excitation beam path consisting of a Glan-Taylor-prism, a $\lambda/2$ retarder waveplate, a $\lambda/4$ retarder waveplate and the streak camera trigger diode, the helium-flow cryostat, the microscope objective, the optics in the detection beam path consisting of a $\lambda/4$ retarder waveplate, a $\lambda/2$ retarder waveplate, a Glan-Taylor-prism and an interference filter, the monochromator and the streak camera. Alternative components for different experimental situations and optional additional elements are described in the text.

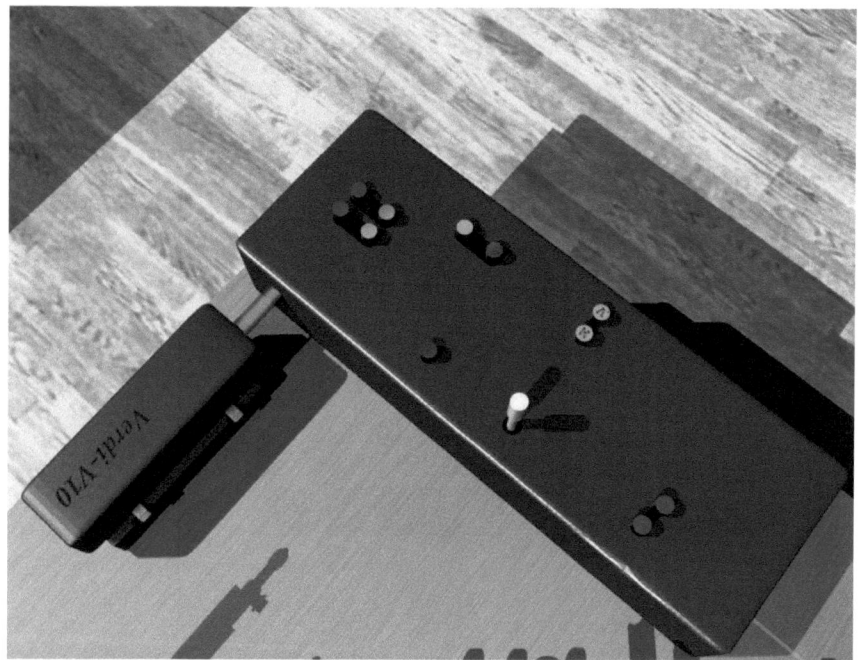

Figure 2.3: Mode-locked Coherent Mira 900-D Ti:sapphire laser system, pumped by a 10 W Nd:YVO$_4$ CW pump laser. The emission is wavelength-tunable between 700 and 980 nm. It is possible to create pulse durations in the femtosecond or picosecond range at a pulse repetition rate of 75.39 MHz. The minimum spectral full width at half maximum of the pulses achievable for picosecond pulses is on the order of 2 meV. At 9 W pumping power the highest reachable Mira output power is roughly 1.4 W at an emission wavelength of 800 nm. At a wavelength of 900 nm the available output intensity reduces to approximately 1 W.

Figure 2.4: Optics in the excitation beam path. A beam splitter directs a small fraction of the laser pulse to the fast streak camera trigger diode. The excitation laser intensity at the sample position can be adjusted by two neutral density gradient filters. In order to minimize aberrations caused by moving the neutral density gradient filters, they are aligned in reverse orientation to each other. A Glan-Taylor-prism is used to create a horizontal beam polarization with an extinction ratio on the order of 10^5. An achromatic $\lambda/2$ and an achromatic $\lambda/4$ retarder waveplate allow to create any linear, circular or elliptical polarization.

Figure 2.5: The microscope objective used to collect the emitted light from a wide range of emission angles. Microscope objectives with numerical apertures of 0.26 (10x objective), 0.42 (50x objective) and 0.5 (100x objective) have been used. Depending on the emission wavelength either microscope objectives with minimized chromatic aberration and focus distance in the near infrared or visible part of the spectrum were used. All microscope objectives are infinity-corrected.

Figure 2.6: The helium-flow cryostat. The sample is mounted on a cold finger inside the cryostat which is in constant contact with helium flowing through the cryostat. The temperature at the sample position can be varied in a range between 6 and 300 K by adjustments of the helium flow rate and by electrical heating. The cryostat is mounted on three micrometer-screw driven translation stages allowing for a course-positioning of the sample location in all three dimensions.

Figure 2.7: The optics in the detection beam path. They consist of an achromatic $\lambda/4$ retarder waveplate, an achromatic $\lambda/2$ retarder waveplate, a Glan-Taylor-prism and an interference filter. The polarization optics allow to perform measurements on a freely chosen polarization component of the emission. The interference filter has a spectral FWHM of approximately 1 nm. Its transmission wavelength can be tuned in a range of approximately 5 nm by turning it.

therefore also the sample position can be course-adjusted by three micrometer-screw operated translation stages.. For both excitation schemes the microscope objectives are used to collect the photoluminescence from the sample. In the case of pumping under normal incidence, the transmitted 92 % are guided towards the detection equipment. The reflected 8 % are not used. It is possible to place another lens behind the microscope objective to perform momentum-space dispersion measurements. For this experiment, the lens is placed a focus distance away from the Fourier plane of the emitted light. The emission is then again going through a set of polarization optics shown in figure 2.7 consisting of an achromatic $\lambda/4$ retarder waveplate, an achromatic $\lambda/2$ retarder waveplate and a Glan-Taylor-prism. This assembly allows to single out a freely chosen linear or circular polarization component of the emitted light. Additionally the emission passes through a narrow bandwidth interference filter with a spectral FWHM of approximately 1 nm, allowing to select a single spectral emission mode if necessary. The interference filter transmission wavelength can be tuned in a range of roughly 5 nm by turning the filter. For recording momentum space dispersion spectra an additional lens can be placed behind the microscope objective. It is used to collimate the Fourier plane image of the emission which carries information about the photon momentum. The emission is then either directed towards the streak camera for time resolved measurements or towards a monochromator (Acton SP-2500i) for spectrally resolved measurements of the real or Fourier plane. The monochromator is shown in figure 2.8. Lenses with a focal length of 150 mm or 200 mm were used to focus the real or Fourier space signal on the entrance slit of the monochromator. It has a focal length of 500 mm and is equipped with a triple grating turret. The mounted gratings exhibit groove densities of 300, 600 or 1200 grooves per millimeter, respectively. All of the gratings are blazed in order to optimize performance for a certain wavelength range. The 300 grooves per mm and the 1200 grooves per millimeter gratings are blazed at 1000 nm. The 600 grooves per mm grating is blazed at 500 nm. The dispersed light shines on a CCD camera mounted in one of the two monochromator exit ports. In order to minimize readout noise and dark counts, the CCD camera is cooled using liquid nitrogen and operated at approximately 180 K. The CCD consists of 1340x400 pixels with a pixel size of 20 μm. Under ideal conditions, the best possible wavelength resolution using this setup is roughly 0.09 nm. The other monochromator exit port can in principle be used to guide the dispersed emission to the streak camera to have a time-resolved and spectrally resolved signal at the same time. However, doing so decreases the achievable temporal resolution to 20 ps, which is not sufficient to perform correlation measurements in the thermal regime. The streak camera is depicted in figure 2.9. A lens with a short focal length of 50 mm is used to focus the signal on the entrance slit. The streak camera is used in synchroscan mode and is synchronized with the mode-locked laser using the signal of the fast trigger diode. Details on the streak camera measurements will be presented in the following section.

2.2.1 The correlation streak camera technique

In a standard streak camera the light pulse to be investigated is projected onto a slit and then focused on the streak tube photocathode. There the photons are converted into an intensity-dependent number of photoelectrons. These are then accelerated towards a micro-channel plate (MCP) using an electrode. On this way they are subject to another pair of sweep electrodes. A

Figure 2.8: The monochromator. It is equipped with a liquid-nitrogen cooled CCD camera operated at approximately 180 K. The monochromator has a focal length of 500 mm and is equipped with a triple grating turret. The mounted gratings exhibit groove densities of 300, 600 or 1200 grooves per millimeter, respectively. All of the gratings are blazed in order to optimize performance for a certain wavelength range. The 300 grooves per mm and the 1200 grooves per millimeter gratings are blazed at 1000 nm. The 600 grooves per mm grating is blazed at 500 nm. The CCD camera consists of 1340x400 pixels with a pixel size of 20 μm. Under ideal conditions, the best possible wavelength resolution using this setup is roughly 0.09 nm.

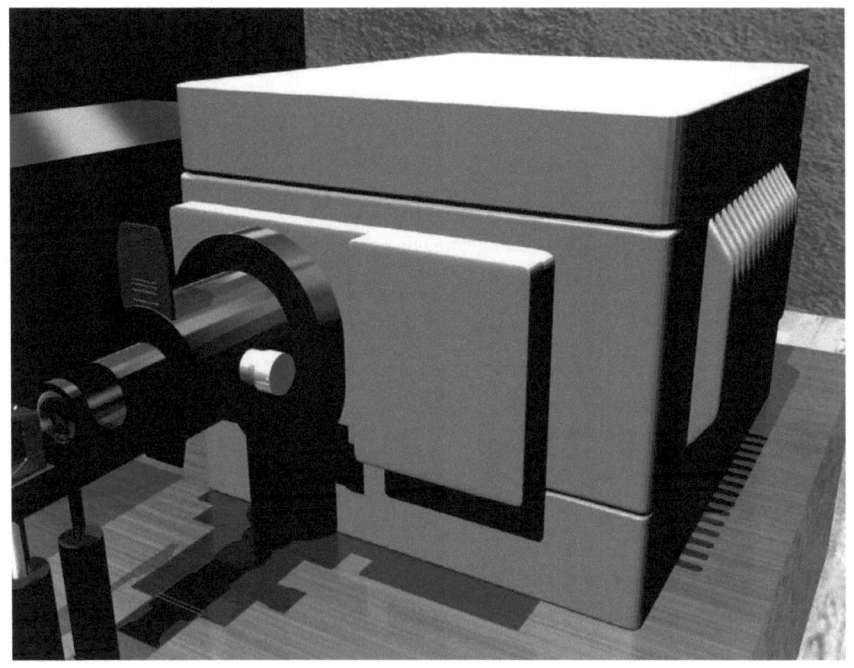

Figure 2.9: The streak camera. A lens with short focal length of 50 mm is used to focus the signal on the entrance slit. The streak camera is synchronized with the mode-locked laser using the signal of the fast trigger diode.

high voltage is applied to these, resulting in a fast vertical sweep of the photoelectrons. Electrons passing the sweep electrons at different times are deflected at different angles and then conducted to the MCP. As the electrons pass the MCP, they are multiplied and hit a phosphor screen. The brightness of the phosphorescence is proportional to the intensity of the incident light and the screen positions in horizontal and vertical directions correspond to the position of the incident light pulses in the horizontal direction and their arrival time, respectively [31]. This phosphorescence image is recorded by a CCD camera.

Using a streak camera for correlation measurements requires further customization. Usually the streak image is recorded by integrating over many repeated streak cycles. However, information about correlations is only present in single pictures and gets washed out by this integration. It is therefore necessary to record images of single signal pulses. This prerequisite significantly reduces the possible data acquisition rate due to the limited readout rates achievable for CCDs. State of the art CCDs with sufficient quantum efficiency for recording single streak pictures can usually be operated at 100 Hz at best. Comparing this value to the excitation laser repetition rate of 75.39 MHz shows that this approach is rather inefficient, using approximately only 1 out of 10^6 pulses. The efficiency can be increasing by utilizing the final degree of freedom left in the streak camera picture: the horizontal position. In common streak camera experiments the horizontal position is used to achieve spectral resolution by placing a monochromator in front of the streak camera. The horizontal position then becomes an indicator for the photon wavelength. For the photon correlation measurements performed in this work only the correlations of a single mode of interest are considered. Therefore it is sufficient to single out this mode. Further spectral resolution is not needed. A sensible approach allowing higher effective readout rates can be realized by placing consecutive signal pulses at several horizontal positions of the screen. Technically this is achieved by adding a pair of horizontal sweep electrodes and applying a second time-dependent voltage to them. The horizontal deflection sweep speed is much slower than the vertical one, which is on the order of the whole screen corresponding to a time-window of 140 ps. A slower sweep speed increases the number of consecutive pulses that can be positioned on one screen. The lower bound of the sweep speed is given by the spatial width of the individual pulses on the screen and the prerequisite that these pulses need to be distinguishable. A reasonable choice of the horizontal sweep speed is given by setting the whole horizontal time-window to 600 ns, allowing one to record 30 to 40 pulses on a single screen. Although the resulting data acquisition rate of approximately 3 kHz is still small compared to the laser repetition rate, it already allows performing measurements in sensible integration times. For reasonable photon count rates taking 10^6 of these single pictures suffices. This corresponds to an acquisition time of 20 to 25 minutes. Advanced CCDs which can be operated at higher frame rates at comparable quantum efficiency might be available within the next few years. In order to take advantage of the improved data acquisition rate those CCDs would offer, it is necessary to customize also the phosphor readout screen. The phosphor afterglow time should not be much shorter than the time required to acquire a single picture, but should be significantly shorter than the waiting time between two pictures to avoid accumulation of afterglow from previous images. The most common phosphor used for streak cameras is P 43 ($Gd_2O_2S:Tb$), which is characterized by a 90% − 10% afterglow decay time of 1 ms. Although that timescale is still short enough for usage with modern CCDs, it is desirable to have faster decay times. We implemented P 46 phosphor ($Y_3Al_5O_{12}:Ce$), which has a characteristic 90% − 10% afterglow decay time of only 300 ns. This shortened decay time comes at the drawback of reduced efficiency. We compensate this effect by adding a second MCP stage, which allows to increase the maximum achievable gain in the streak tube. The final requirement is single photon sensitivity. This has already been demonstrated in single photon counting mode [32].

In order to determine the photon correlation function as defined by equation 1.56, it is necessary to measure the numerator and denominator separately. The numerator corresponds to the detected number of photon pairs. The denominator equals the expected number of photon pairs for the same mean photon count rates, but statistically independent emission. While the first quantity can be evaluated by counting the number of photon pairs in all the recorded single images, the latter quantity is available in the integrated picture, where all correlations have vanished. These two images are compared in figure 2.10. The upper panel shows the integrated picture. This image contains information about the mean photon count per image and pixel. It is also used to evaluate the position of the single streaks on the screen. The lower panel shows a single image. Each detected photon is sorted to a streak corresponding to its horizontal position and to a time bin depending on its vertical position. These time bins define the vertical span of pixels which are considered to have arrived simultaneously. Their size defines the effective temporal resolution, which can be chosen after the experimental data has already been taken. Now all photon pairs inside the pulses are counted and also sorted to bins depending on the arrival time of the first photon and the delay between the two photons. Now $g^{(2)}(t,\tau)$ can be determined by counting all the photon pairs in bins corresponding to times t and $t+\tau$ and dividing this number by the product of the mean photon count rates at the corresponding times and averaging this value over all streaks. An averaged $g^{(2)}(\tau)$ without t-dependence can be evaluated by calculating the weighted average of all $g^{(2)}(t,\tau)$ for a certain

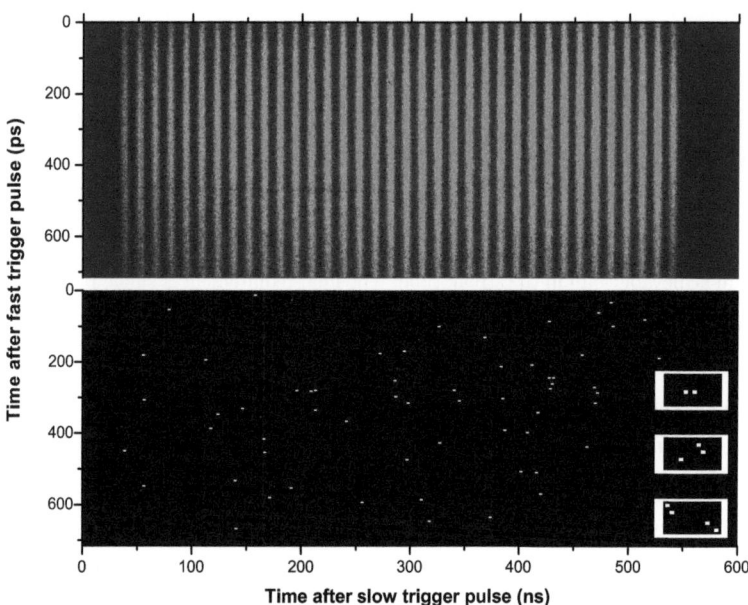

Figure 2.10: Comparison of an integrated and a single streak camera image with vertical and horizontal time windows of 716 ps and 600 ns, respectively. The upper panel shows an integrated picture consisting of 100000 single pictures. 36 consecutive pulses are shown. This image contains information about the mean photon count rates at each pixel. The lower panel shows one single streak camera image. White dots correspond to photon detections. Insets show enlarged examples of two-, three- and four-photon coincidences.

τ:

$$g^{(2)}(\tau) = \frac{\sum_t \langle : \hat{n}(t)\hat{n}(t+\tau) : \rangle}{\sum_t \langle \hat{n}(t)\rangle\langle \hat{n}(t+\tau)\rangle}. \tag{2.1}$$

This approach can be generalized to arbitrary higher orders n by analyzing n-photon coincidences instead of two-photon coincidences.

2.2.2 Characterization of Streak camera performance: time-integrated measurements

Whether the measured $g^{(2)}$ represents the real correlation function or is distorted by the measurement apparatus is mainly determined by two figures of merit characterizing the detector used: temporal resolution and dark count rate. When considering the equal time correlation function $g^{(2)}(\tau = 0)$ the length of the time window during which detections are considered to be simultaneous is determined by the temporal resolution t_r of the experimental setup. If the dynamics of $g^{(2)}(\tau)$ varies strongly within t_r, the measured value of $g^{(2)}(0)$ will be the averaged value of $g^{(2)}(\tau)$ within this time window. As $g^{(2)}(\tau)$ usually decays towards unity, the measured value of $g^{(2)}(0)$ will also be shifted towards unity compared to the real value. The timescale of this decay is given by the coherence time τ_c. Therefore the ratio of these two characteristic timescales gives a good impression whether a detector is suitable to measure the correlation function of a signal with known coherence time or not. In detail the measured $g^{(2)}$ will be given by the convolution of the real expression with a Gaussian function of width $2\sigma = t_{IRF}$ [33] representing the instrument response function. For thermal light of an inhomogeneously broadened PL spectrum the real $g^{(2)}(\tau)$ is given by [34]

$$g^{(2)}(\tau) = 1 + \exp\left(\frac{|\tau|^2}{\tau_c^2}\right) \tag{2.2}$$

and the convolution reads

$$g^{(2)}_{IRF}(\tau) = \frac{1}{\sqrt{2\pi\sigma^2}} \int_{-\infty}^{\infty} d\tau' g^{(2)}(\tau - \tau') \exp\left(\frac{-\tau'^2}{2\sigma^2}\right). \tag{2.3}$$

As can bee seen in figure 2.11 the measured and the real correlation functions are in good agreement when t_{IRF} is much shorter than τ_c. If the timescales become comparable, the measured values start to differ from the real ones, leading to underestimates of 3 % if τ_c is twice as large as t_{IRF} and 10 % if both are equal. Decreasing the temporal resolution further, the underestimation becomes critical, causing drastic deviations already if t_{IRF} is twice as large as τ_c. Therefore the real temporal resolution of the streak camera setup is of major interest. The

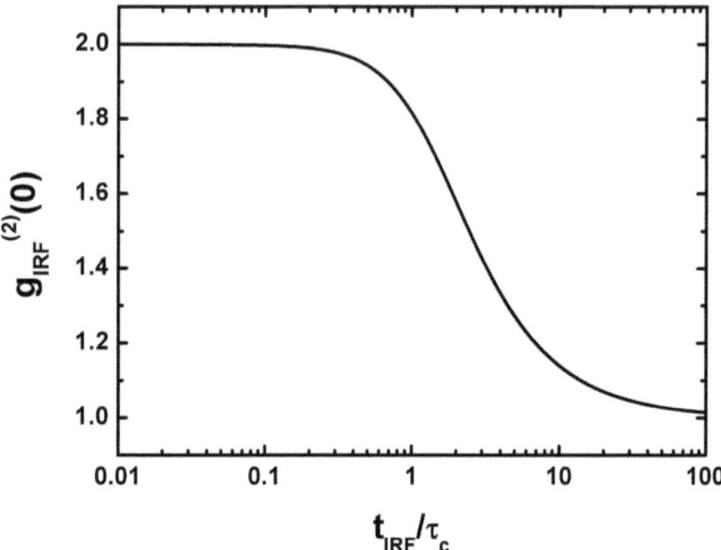

Figure 2.11: Expected measured correlation function $g_{IRF}^{(2)}(0)$ as a function of the ratio $\frac{t_{IRF}}{\tau_c}$ of the detector temporal resolution and the signal coherence time for thermal light with $g^{(2)}(0) = 2$.

theoretical limit is given by the number of pixels on the readout CCD. Using the smallest time window present a whole screen length consisting of 480 pixels corresponds to a time window of 136 ps. Accordingly, the optimal t_{IRF} possible is 283.3 fs. However, the real t_{IRF} is caused by several jitter sources, of which timing jitter of the trigger pulse, emission jitter of the photocathode and and broadening caused by the MCPs are the most important. Typically t_{IRF} is on the order of 2 ps. More exact measurements of the jitter are possible, but require background noise to be considered, too.

The effect of background noise is comparable to the presence of a second mode, which is uncorrelated with the first one. The equal time correlation function for two noninterfering modes A and B with mean intensities I_A and I_B then depends on the relative count rates $R_A = \frac{I_A}{I_A+I_B}$ and $R_B = \frac{I_B}{I_A+I_B}$:

$$g_n^{(2)}(0) = R_A^2 g_A^{(2)}(0) + R_B^2 g_B^{(2)}(0) + 2R_A R_B. \qquad (2.4)$$

In the following A is considered to be a thermal mode with $g_A^{(2)}(0) = 2$ and B represents Poissonian noise with $g_B^{(2)}(0) = 1$. The addition of noise will also cause an underestimation of $g^{(2)}(0)$ below the real value as shown in figure 2.12. As expected, without noise the measured value

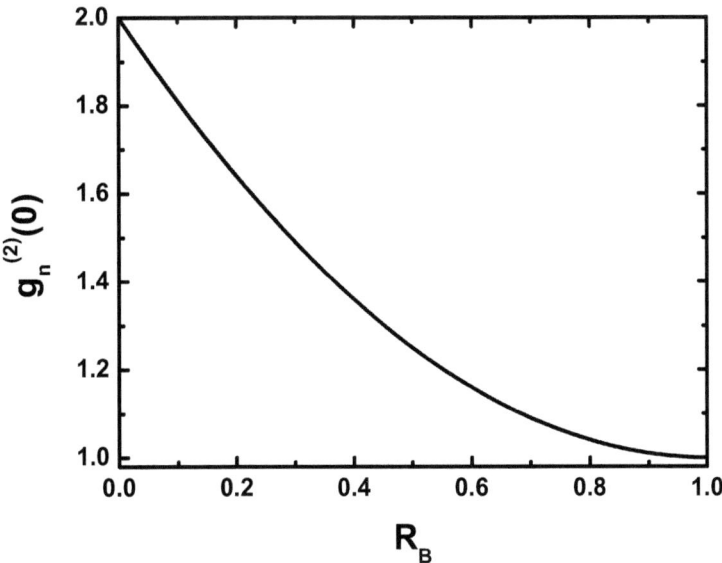

Figure 2.12: Expected measured correlation function $g_n^2(0)$ for a thermal mode as a function of the relative noise intensity R_B.

reaches the real value of two, while large noise fractions reduce the measured value towards unity. For a noise value of 15 % the measured value of $g^{(2)}$ is approximately 15 % smaller than the real value. For smaller R_B this dependence is approximately linear. For larger values of R_B the underestimation becomes severe, so it is of importance to reach a high signal-to-noise ratio.

Another considerable source of deviations are systematically erroneous photon detection events. The most prominent reason for such events lies in the built-in streak camera photon reconstruction algorithm. Any detected photon will not fill exactly one pixel, but will have a size of several pixels. The photon reconstruction algorithm reconstructs the photon position on the screen by finding the center of gravity of the intensity measured at adjacent pixels. Local inhomogeneities of the streak camera screen can fool the photon reconstruction routine to consider one single photon as two or vice versa. However, this problem can easily be detected by analysis of the distribution of pixel distances between two identified photons. Incorrect photon reconstruction will result in a distribution with very high pair count rates for pixel distances shorter than the radius R_{photon} of the photon size on the screen, which abruptly falls off steplike at distances, which are larger than R_{photon}. To fix this problem, it is possible to introduce artificial dead pixels. If all photons detected within a distance of R of another photon are disregarded, any erroneous photon counts due to reconstruction problems are eliminated. However in this way also real photon detections are disregarded. Effectively, the detector size for detecting another photon after the first one went down, which simply modifies the normalization by a factor of $\frac{wA_{red}}{A}$ where A is the size of a bin on the screen, A_{red} is the reduced size after introducing artificial dead pixels and w is a weighting factor, which is needed, if the mean number of photon counts is not distributed equally along the width of a bin. If artificial dead pixels are used in measurements of higher order correlation functions, it is necessary to consider the decreasing effective detector size inside a time bin after each photon detection, accordingly. However, determining this weighting factor for spatially inhomogeneous signals is nontrivial and the artificial dead detector area usually corresponds to a time span of 1.5 ps at worst and can be reconstructed if enough data points at larger delay are known. An example will be given in the following section.

Both t_{IRF} and R_B can be determined by measuring a signal of well known $g^{(2)}$ and short duration. If the signal duration becomes comparable to the jitter width, a significant distortion of the measured $g^{(2)}$ can be expected. In this situation the major contribution to the fluctuations of the photon pair detections at a fixed screen positions is caused by the jitter placing a slightly different portion of the jitter at that screen position for each single image taken. For short signals this fluctuation in the momentary intensity in each single image is large compared to the intrinsic fluctuations due to photon statistics. This opens up the possibility to determine the magnitude of the timing jitter by a simple model describing the measured $g_j^{(2)}$ in the presence of jitter. N_{rep} denotes the number of pictures taken and p is an index identifying a single image. In analogy the momentary value of the jitter for a single image is given by j_p. j_p is normal-distributed around zero. A constant noise background photon count rate is given by r_n. $A_p(t)$ describes the total photon number of a signal pulse in a single picture, which follows a probability distribution depending on the kind of light source considered: a Poissonian distribution for coherent light or a Bose-Einstein distribution for thermal light. Finally, $S(t)$ gives the pure temporal shape of the signal without any influences of jitter or noise, with the area under the curve normalized to unity. The resulting $g_j^{(2)}$ depends strongly on the discrete convolution of $S(t)$ and the Gaussian jitter. As it might also depend on the chosen temporal resolution for thermal light, a discrete number of values for t is used:

$$g_j^{(2)}(\tau) = \frac{\frac{1}{N_{rep}}\sum_{p=0}^{N_{rep}}\sum_t [(r_n + A_p S(t+j_p))(r_n + A_p S(t+\tau+j_p))]}{\frac{1}{N_{rep}}\sum_{p=0}^{N_{rep}}\sum_t [r_n + A_p S(t+j_p)] \frac{1}{N_{rep}}\sum_{p=0}^{N_{rep}}\sum_t [r_n + A_p S(t+\tau+j_p)]}. \qquad (2.5)$$

This is the jitter-broadened photon pair count rate divided by the product of two broadened single photon count rates. Accordingly the numerator will not be broadened as much as the denominator.

As an experimental test signal, the excitation laser pulse was used, which has Gaussian shape, a well known standard deviation of $\sigma = 1.42\,\text{ps}$ and $g^{(2)}(\tau) = 1$ for all τ inside the pulse. The usage of a fully second-order coherent light source makes it possible to use a simplified treatment instead of equation 2.5 as the photon pair count rate will factorize into the product of the mean photon count rates. In this case it is sufficient to calculate the convolution of the continuous pulse shapes with the jitter distribution of standard deviation J:

$$g_j^{(2)}(t,\tau) = \frac{\int (S(\tau_2) + r_n)(S(\tau + \tau_2) + r_n) J(t - \tau_2)\, d\tau_2}{(\int (S(\tau_2) + r_n) J(t - \tau_2)\, d\tau_2)(\int (S(\tau + \tau_2) + r_n) J(t - \tau_2)\, d\tau_2)}. \qquad (2.6)$$

The single photon count rates calculated in the denominator are convolutions of two Gaussians and can be described as another Gaussian $S_j(t)$ with a modified standard deviation of $W_j = \sqrt{W^2 + J^2}$. The time-averaged $g_j^{(2)}(\tau)$ is then given by:

$$g_j^{(2)}(\tau) = \frac{\int \int (S(\tau_2) + r_n)(S(\tau + \tau_2) + r_n) J(t - \tau_2)\, d\tau_2\, dt}{\int (S_j(t) + r_n)(S_j(t + \tau) + r_n)\, dt}. \qquad (2.7)$$

The product of the jitter-broadened single-photon count rates will usually be a broader distribution than the jitter broadened photon pair count rate. Therefore the streak camera can be described as a jitter sensitive autocorrelator. An example of this behavior is shown in figure 2.13 for $J = W = 1.42\,\text{ps}$ and $r_n = 0.0033$. The jitter-broadened photon pair detection rate depending on the delay τ given by the red line shows a narrower distribution than the expected photon pair detection rate as calculated by the single photon detection rates (blue line). This deviation can be explained by looking at the τ-dependence of the photon pair count rates inside a single picture. Inside single images the t-integrated τ-dependence of the photon count rates is not affected by jitter. However, the mean photon count rates are broadened by jitter and lead to the significant shape of $g_j^{(2)}(\tau)$ seen in figure 2.13. It consists of three characteristic regions. For short delays the pair count rate exceeds the expected value because its width is mainly

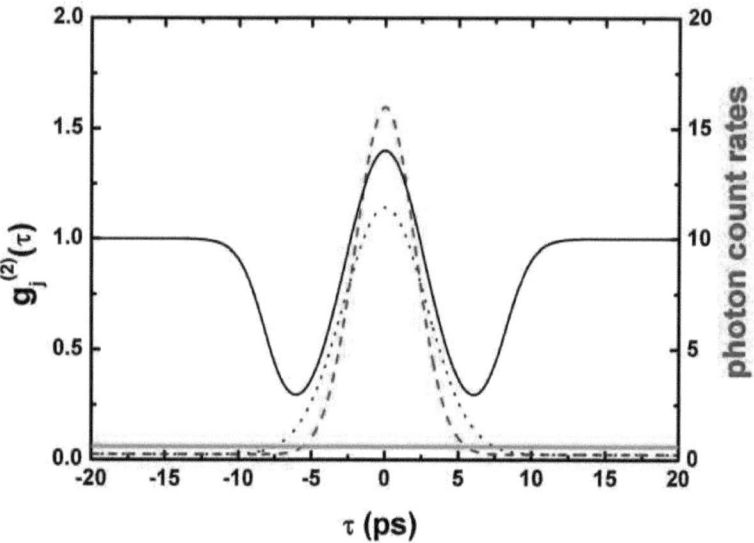

Figure 2.13: Comparison of the photon pair counting rate (dashed line) to the product of the jitter-broadened mean photon count rates (dotted line) at a time delay of τ for $J = W = 1.42\,\text{ps}$ and single photon noise count rate $r_n = 0.0033$ (bright solid line). The solid black line gives the corresponding $g_j^{(2)}(\tau)$.

determined by the pulse width, while the mean photon count rate is jitter broadened. This overshoot is a good measure of J. However, the faster decay of the pair count rate compared to the expectation value causes $g_j^{(2)}(\tau)$ to decrease below unity at some finite τ. This decay is mainly determined by the pulse width and therefore a good measure of W. Finally $g_j^{(2)}(\tau)$ increases towards unity again for large τ. This increase is not caused by the signal, but by the intensity of the background noise denoted by the solid green line becoming larger than the signal intensity and is a good measure of r_n. It is possible to give an estimate of these quantities by fitting 2.7 to a given autocorrelation trace. Some fits to the autocorrelation of the excitation laser are shown in figure 2.14 for several settings of the MCP gain and the photon counting threshold. The choice of MCP gain and threshold settings is crucial because the double-MCP stage we use creates a broad distribution in the actual number of photoelectrons emitted per detected photon and their spatial distribution. The left panel shows experimental results for

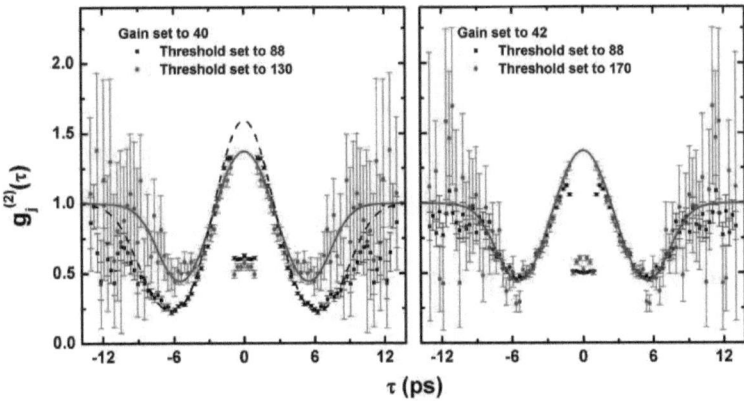

Figure 2.14: Jitter induced autocorrelation trace of a laser pulse with FWHM of 3.34 ps. for gain settings of 40 (left panel) and 42 (right panel). Experimental results for low and high threshold settings are shown as squares and dots, respectively. Lines represent fits according to equation 2.7. The solid lines correspond to fits for $J = 1.38$ ps, $W = 1.42$ ps and $r_n = 0.0033$. The dashed line corresponds to a fit for $J = 1.81$ ps, $W = 1.42$ ps and $r_n = 0.0015$.

a MCP gain of 40 and a low (black squares) and high (red dots) photon counting threshold, respectively. At first it is obvious that the results for τ below 1.3 ps are far too small. This is a result of introducing a 3x3 dead pixel area around each detected photon. The experimental values are not corrected for the smaller effective detector area. However, the necessary correction factor could be obtained from this data. Fits to the experimental results giving two different sets of parameters are shown as solid black and red lines. The black line is a fit to the low threshold results with $W = 1.42$ ps, $J = 1.81$ ps and $r_n = 0.0015$, while the red line corresponds to $W = 1.42$ ps, $J = 1.38$ ps and $r_n = 0.0033$. As normalized probability distributions are used to fit $g_j^{(2)}(\tau)$, r_n is inversely proportional to the signal-to-noise ratio. As can be seen a low threshold setting gives a good signal-to-noise ratio, but also some underestimation of $g_j^{(2)}(\tau)$ for delays which correspond to streak camera positions slightly beyond the artificial dead area. This is an artifact of the photon reconstruction algorithm. For small thresholds R_{photon} can become large. If two photons overlap at the phosphorus screen, the photon reconstruction algorithm will sometimes detect them as only one photon causing an underestimate of $g_j^{(2)}(\tau)$.

Usually this effect is countered by introducing artificial dead pixels, but if R_{photon} is larger than the dead area artifacts remain. These artifacts also explain the large value of J seen in the black autocorrelation trace. The variation in the center of gravity of the detected intensity will be very large for a case when two photons are erroneously counted as one. For larger delays the black trace matches the experimental data well. Also even the center of gravity of single photon detections might not be well defined for small threshold settings. The experimental data for larger photon counting threshold settings does not show this underestimation of $g_j^{(2)}(\tau)$ and reveals a significantly smaller jitter, but suffers from a slightly worse signal-to-noise ratio. Therefore this setting should be chosen for strong signals. The same signal at increased MCP gain is shown in the right panel. Black Squares and red dots again denote low and high threshold settings, respectively. The solid red line is a fit using the same parameters as the solid red line in the left panel. As can be seen, neither the dark count rate, nor the timing jitter change at small changes of the MCP gain. The fit matches the experimental data for both threshold settings well, although the underestimation of $g_j^{(2)}(\tau)$ is even more pronounced for small threshold settings and small τ. The jitter does, however, not show the drastic increase seen for small threshold setting at lower gain. This might indicate that due to the larger gain the photon spot shape on the phosphor screen becomes more homogeneous. Therefore the photon reconstruction algorithm estimates the real center of gravity with higher precision. These results demonstrate that the streak camera correlation principle is insensitive to the choice of streak camera settings within a suitable range.

Especially the value of $g_j^{(2)}(0)$ is a remarkable quantity. It depends strongly on the ratio of J and W and is therefore important to distinguish between real and jitter-induced correlations for short signals. As shown in figure 2.15 the increased simultaneous photon detection probability due to jitter influence for second and third order goes as:

$$g_j^{(2)}(0) = g^{(2)}(0)\sqrt{1+\left(\frac{J}{W}\right)^2} \tag{2.8}$$

$$g_j^{(3)}(0) = g^{(3)}(0)\left(1+\left(\frac{J}{W}\right)^2\right), \tag{2.9}$$

if the influence of r_n can be neglected. Significant deviations between $g_j^{(2)}(0)$ and $g^{(2)}(0)$ occur if $\frac{J}{W}$ is larger than 0.25. Assuming that J takes on a value of 1.38 ps as determined above for each experimental setting, the measured $g^{(2)}(0)$ should be corrected for jitter effects for signals with FWHM smaller than approximately 10 ps. As the jitter-broadened pulse width is automatically measured in the integrated intensity, all quantities needed to perform the corrections for t-integrated measurements are well known.

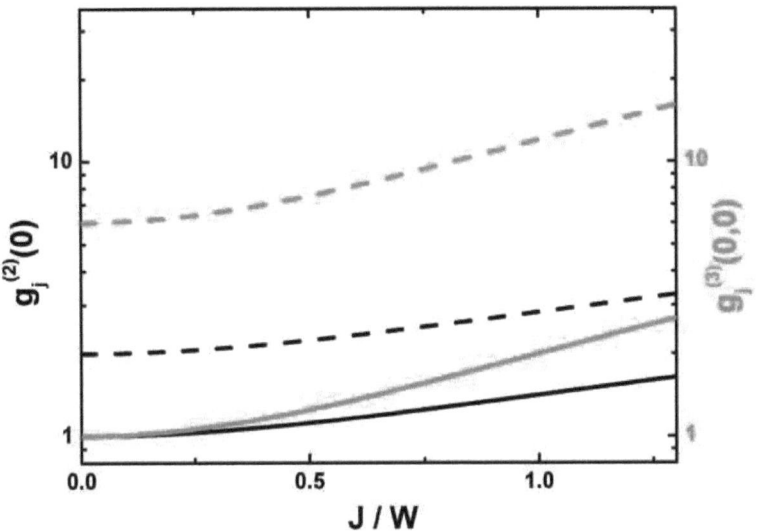

Figure 2.15: Simulated jitter dependence of the second-order (dark line) and third-order (bright line) equal-time correlation functions depending on the ratio between J and W for coherent (solid lines) and thermal (dashed lines) light. Results are shown on a logarithmic scale.

2.2.3 Characterization of Streak camera performance: time-resolved measurements

Up to now mainly intensity correlation measurements without explicit t-dependence have been considered. These results describe the statistical properties of the signal pulse as a whole. If one is interested in the pulse dynamics, this point of view is not sufficient. The photon statistics are not necessarily the same at different times inside the signal pulse. Especially for systems which show a lasing transition, the time resolved correlation function $g^{(2)}(t,\tau)$ can give deeper insight into the relevant dynamics. Although it is in principle possible to perform complete two-dimensional mapping of this function in t and τ, the most relevant part is the time-resolved equal-time intensity correlation function $g^{(2)}(t,0)$. It allows to follow the build-up and breakdown of coherence inside pulses. Although this quantity can in principle be easily achieved by comparing the photon pair count rate inside a time bin to the square of the single photon count rate of the same bin, the necessary corrections due to jitter are more complicated than in the time-integrated case, mainly because the time integrated photon pair count rate per pulse is not affected by jitter, while the time resolved pair count rate at a certain screen position is. For Gaussian pulse shapes a solution of $g_j^{(2)}(t,0)$ can be found from equation 2.6. The origin of the t-axis is in the following assumed to be the time of arrival of the pulse peak. The measured intensity correlation at this peak position $g_j^{(2)}(0,0)$ is a good indicator of how much the real correlation function is masked by the jitter. An analytical solution yields

$$g_j^{(2)}(0,0) = g^{(2)}(0,0) \frac{r_n^2 + r_n\sqrt{\frac{2}{\pi}}\frac{1}{W_j} + \frac{1}{2\pi WB}}{(r_n + \frac{1}{\sqrt{2\pi}W_j})^2}, \qquad (2.10)$$

where $B = \sqrt{2J^2 + W^2}$. This function is depicted in figure 2.16. The jitter-dependence bears some similarities to $g_j^{(2)}(0)$. This is not surprising as the latter function is the average over all $g_j^{(2)}(t,0)$ weighted by the squared mean photon count at time t. $g_j^{(2)}(0,0)$ is the second order correlation function at the pulse peak position and will enter into the averaged value with the largest weight. More interesting is the complete time dependence of the jitter-dependent second order correlation function. Although it is in general more complicated, an analytical form can be given for Gaussian pulse shapes:

$$g_j^{(2)}(t,0) = \frac{g^{(2)}(t,0)}{\left(r_n + \frac{e^{-\frac{t^2}{2W_j^2}}}{\sqrt{2\pi}W_j}\right)^2} \left(r_n^2 + \frac{r_n e^{-\frac{t^2}{2W_j^2}}}{\sqrt{2\pi}W_j} + \frac{1}{2}\left(\frac{e^{-\frac{t^2}{B^2}}}{\pi WB} + \frac{r_n\sqrt{\frac{2}{\pi}}e^{-\frac{t^2}{2W_j^2}}}{W_j} \right) \right). \qquad (2.11)$$

This function is shown in figure 2.17 for values of $J = 1.42\,\text{ps}$ and a constant noise background of $N = 0.0033$ for three different values of W. From the enhanced value at the peak position already given by equation 2.10 the increase in $g_j^{(2)}(t,0)$ becomes even larger in a narrow range around the origin, which tends to broaden with increased W. The reason for this effect lies in the relevant effect of large-jitter events on pair detections at pulse positions with small mean intensity. In these regions the mean intensity will most likely occur due to few large shifts of the pulse peak position inside single recorded pictures. For these few pulses the affected regions will be subject to a rather high intensity, but there will be almost no intensity at these positions in other single recorded pictures. These occurrences of large momentary jitter are rare events and have different effect on regions with low and high intensity. For regions with high mean intensity they cause one single frame of rather low intensity which does not contribute much to the average intensity at that position. For regions with low mean intensity instead, these rare events cause a large portion of the mean intensity. This corresponds to large shot-by-shot photon number fluctuations and causes the significant overshoot seen in $g_j^{(2)}(t,0)$ away from

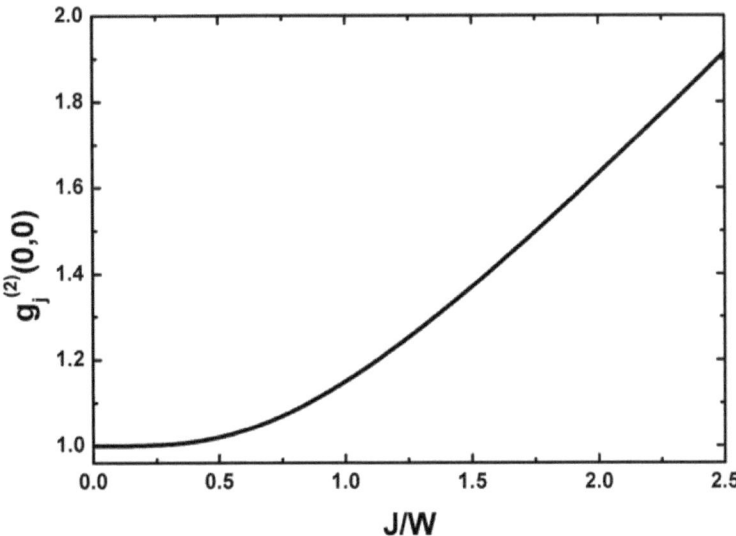

Figure 2.16: Simulated jitter dependence of the second-order correlation function at the position of the pulse peak.

the origin. In regions with even lower mean intensity, the contribution of the noise becomes the dominant source of the mean intensity and even the rate of large jitter events shifting intensity to these positions vanishes. In this region $g_j^{(2)}(t,0)$ decays back to the uncorrelated value of 1. This explains also the broadening of the overshoot region as also the onset of the noise-dominated pulse region shifts to larger times with increased W.

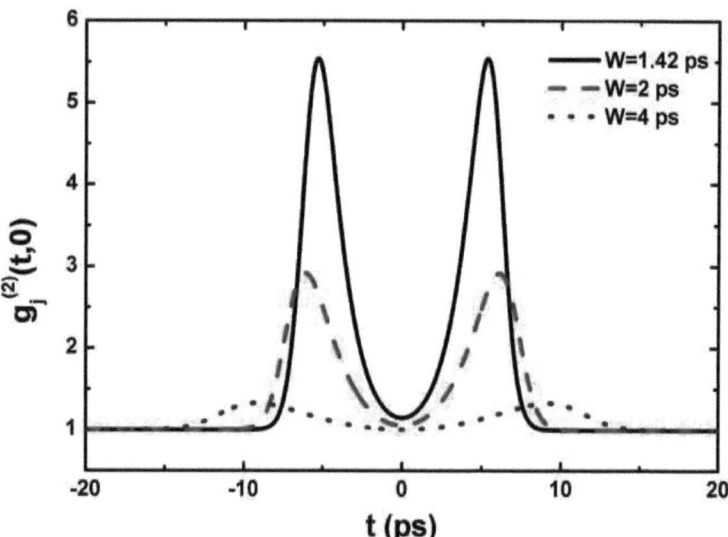

Figure 2.17: Effect of jitter on $g_j^{(2)}(t,0)$ for $J = 1.42$ ps and a constant noise background of $N = 0.0033$. Solid, dashed and dotted lines give the simulated results for values of $W = 1.42$, 2 and 4 ps, respectively.

Chapter 3

Quantum Dot VCSELs

Efforts to realize lasers with extremely small lasing threshold led to the development of vertical-cavity surface-emitting lasers (VCSELs). They consist of a semiconductor nanostructure - either quantum dots or a quantum well - embedded as active medium inside a microcavity in the weak coupling regime. In general the efficiency of lasers depends on several characteristic quantities. As population inversion needs to be created inside the whole active medium, small volumes of the active medium decrease the threshold excitation density significantly. The drawback of small active volumes manifests in small gain per pass which results in the need to achieve high Q-factors in order to establish long photon lifetimes inside the microcavity. Also the β-factor

$$\beta = \frac{\tau_l^{-1}}{\tau_{sp}^{-1}} = \frac{\tau_l^{-1}}{\tau_l^{-1} + \tau_{nl}^{-1}} \qquad (3.1)$$

defining the spontaneous emission rate into a lasing mode τ_l divided by the total spontaneous emission rate into all modes τ_{sp} including also nonradiative recombination processes, has large impact on the lasing threshold [35, 36]. This can be seen nicely by comparing input-output curves for lasers with different β-factors as shown in figure 3.1. A significant shift of the threshold pump rate towards lower thresholds occurs with increasing β-factor which is accompanied by an decrease in the step-like jump of the mean emitted intensity in the threshold region. This decrease is a consequence of the reduction of losses due to nonradiative recombination processes and spontaneous emission into nonlasing modes for high-β-lasers. For atom laser systems this jump scales with β^{-1}. The case of $\beta = 1$ is the so-called thresholdless laser where the lasing transition cannot be identified by the input-output (IO) curve anymore and lasing operation is possible for extremely low excitation powers. Common semiconductor or gas lasers show β-factors on the order of 10^{-7} to 10^{-5} [37] only. Therefore, considerable efforts were devoted to improvements of the β-factor. Two main strategies were proposed: Increasing the spontaneous radiative decay rate in order to minimize the effects of nonradiative recombination channels

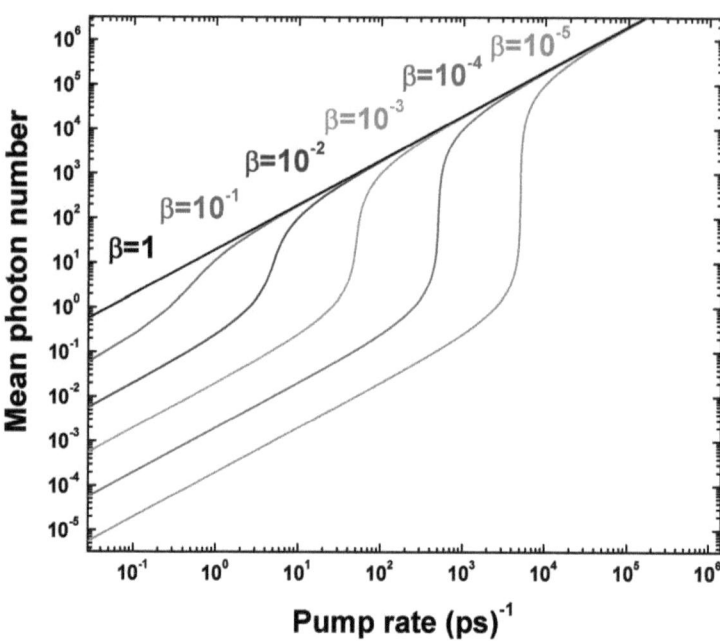

Figure 3.1: Calculated input-output curves for lasers with varying β-factor.

and decreasing the number of photon modes an excited emitter can radiatively decay to. While the latter approach has already been discussed in chapter 1.2.2, the first approach has not been considered in detail yet. In the weak-coupling regime the linewidth $\Delta\lambda_E$ of the emitter inside the microcavity is usually small compared to the spectral width $\Delta\lambda_C$ of the cavity mode. Then the emitter couples to a continuum of modes allowing to describe the emission dynamics by Fermi's golden rule. Due to the tailored optical mode density inside a microcavity, the spontaneous emission rate will be different from the free space emission rate τ_0^{-1} and will in particular depend on the position $\vec{r}$ of the emitter inside the cavity and the detuning between the emitter and the cavity mode as follows [38]:

$$\frac{\tau_0}{\tau} = \frac{2}{3} F_P \frac{|E(\vec{r})|^2}{|E_{max}|^2} \frac{\Delta\lambda_C^2}{\Delta\lambda_C^2 + 4(\lambda_C - \lambda_E)^2} + \alpha \tag{3.2}$$

for an electric field distribution $E(\vec{r})$ with maximum amplitude E_{max}. Here α is the emission rate into leaky modes and the Purcell factor [39]

$$F_P = \frac{3}{4\pi^2} Q \frac{\lambda_C^3}{n_{GaAs}^3 V_m} \quad (3.3)$$

is a figure of merit for a bare microcavity with mode volume V_m, independent of the emitter. Depending on the detuning of the emitter and the cavity, the spontaneous emission rate into the cavity is enhanced or suppressed [40]. For planar cavities F_P is generally of the order of unity [14], but it can reach values around 30 for high-Q micropillars [41]. As a result the radiative decay rate strongly exceeds the decay rates of the nonradiative decay channels, allowing for low-threshold lasing.

In this regime the lasing threshold already becomes smeared out and less well defined. Further complications in deriving correct conclusions arise from semiconductor specific properties of the IO-curve. Saturation effects due to Pauli blocking, different relaxation times for electrons and holes and modified spontaneous emission terms cause significant differences between the IO curves of atomic lasers and semiconductor lasers [35]. One of the most important differences is found in the β-dependence of the kink in the IO curve: for QD lasers this kink does not scale with β^{-1} anymore, but shows nontrivial behavior. Therefore, it has been suggested that the combination of the IO curve and an analysis of corresponding photon statistics gives a more accurate interpretation of the nature of the emitted light [2] for high-β lasers.

3.1 QD VCSEL Samples

Three different QD micropillar lasers with different characteristics were studied. All of them have nominally cylindrical shape. One sample based on a II-VI material system and two III-V material system based samples were used. The II-VI based sample was grown by molecular beam epitaxy. The distributed Bragg reflectors consist of 15 upper and 18 bottom layers in which $ZnS_{0.06}Se_{0.94}$ (48 nm) layers were used as high-index material and a 25.5-period MgS (1.7 nm)ZnCdSe (0.6 nm) superlattice was used as low-index material. The central λ cavity contains a single sheet of CdSe/ZnSe quantum dots with an approximate density of $\sim 5 \times 10^{10}$ cm^{-2}. A pillar with 1.5 μm diameter was used for the measurements. The cavity quality factor was estimated to be ~ 1850.

The III-V micropillar samples were grown by molecular beam epitaxy on a GaAs substrate. The distributed Bragg reflectors consist of 20 upper and 23 lower alternating layers of AlAs (79 nm)/GaAs (67 nm) $\lambda/4$ pairs for the low-Q micropillar and 26 upper and 33 lower alternating layers of AlAs (74 nm)/GaAs (68 nm) $\lambda/4$ pairs for the high-Q micropillar. The central λ cavity contains one layer of self-assembled InGaAs quantum dots with a density of $\sim 3 \times 10^{10}\,\mathrm{cm}^{-2}$ in the low-Q case and one layer of self-assembled AlGaInAs quantum dots with a density of $\sim 6 \times 10^9\,\mathrm{cm}^{-2}$ in the high-Q case, from which cavities with diameters of several micrometres were fabricated by means of high-resolution electron beam lithography and plasma-induced reactive ion etching. Micropillars with diameters of 5 μm (low-Q) and 8 μm (high-Q) were used for the experiments. The Q-factors are deduced from the linewidths of the fundamental modes seen in the modal spectra. Taking the finite resolution of the spectrometer into account, they are estimated as 9000 and 19000, respectively. A typical mode spectrum taken at high excitation power under non-resonant, pulsed optical excitation of 1 mW for the low-Q III-V micropillar is shown in figure 3.2. The fundamental mode is located at 893.53 nm. The first excited mode

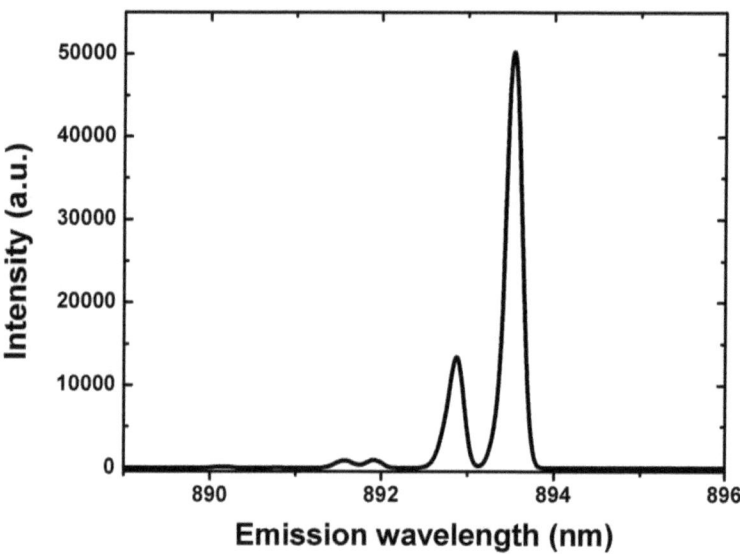

Figure 3.2: Longitudinal emission mode spectrum of the low-Q micropillar laser. The fundamental mode is twofold degenerate and shows emission at 893.53 nm. The polarization splitting cannot be resolved. The first four excited modes can also be seen.

can be seen at an emission wavelength of 892.85 nm. This difference is large enough to single out the fundamental mode using an interference filter with 1 nm spectral width.

3.2 Correlation Measurements on QD VCSELs

The main aim of the measurements was to investigate the basic emission properties of the three considered samples under varying excitation power. At low excitation densities, a broad emission peak from the QD ensemble is seen, superimposed by a series of narrow high intensity peaks marking the microcavity modes. With increasing excitation density, the integrated fundamental mode intensities of all pillars show a characteristic slope change in double-logarithmic plots as shown in the lower panel of figure 3.3. This nonlinearity marks the onset of stimulated emis-

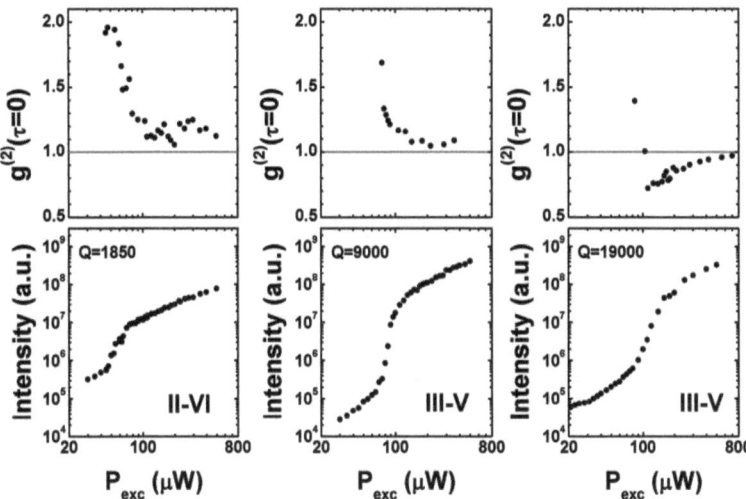

Figure 3.3: $g^{(2)}(0)$ (upper panel) and corresponding input-output curves (lower panel) for three different QD lasers. The left column shows results for a 1.5 μm diameter II-VI cavity. The other columns show results for III-V cavities of diameters 5 μm (middle column) and 8 μm (right column). Solid lines in the upper panel denote the coherent limit.

sion in the microlaser structures. The nonlinear region is apparently broadened over a range of excitation densities. For all three samples the width in excitation powers of this broadened region roughly equals the excitation power at its onset which complicates the definition of a well defined lasing threshold. Determining the β-factor is another nontrivial task as the samples operate in a regime where the kink in the IO curve does not scale with $β^{-1}$. Theoretical analysis reveals β-factors on the order of ∼0.1 for the III-V cavities and a slightly higher β-factor on the order of ~0.13 for the II-VI cavity [42]. Complementary measurements of the equal-time second-order correlation function using the experimental setup presented in chapter 2 allow for a more detailed analysis of the properties of the emitted light. Results are shown in the upper panel of figure 3.3. At high excitation densities far above the lasing threshold all samples show lasing emission identified by values of $g^{(2)}(0)$ of approximately 1 which are clear evidence for the Poissonian nature of the underlying photon statistics. The low-Q III-V and the II-VI sample are still subject to some small excess fluctuations which manifest in values of $g^{(2)}(0) = 1.1$ and 1.2, respectively. The origin of this small amount of excess noise is not completely clear. Possible reasons include relevant contributions of spontaneous emission from early and late times in the pulse and efficient cavity feeding effects [43, 44]. Further studies on the influence of spontaneous emission at different emission times inside the pulse will be described later on, in chapter 3.3. Below threshold the behavior is rather different for the three samples. None of the samples shows a saturation of $g^{(2)}(0)$ at a value of 2 as would be expected for a classical low-β laser. For the II-VI cavity $g^{(2)}(0)$ saturates for low excitation powers at a value of $\approx 1.9 - 1.95$. The small difference from the expected value for a low-β laser is a manifestation of a limited number of emitters. A detailed treatment shows that thermal radiation emitted by a system consisting of a fixed number N of emitters shows an emitter-number dependent second-order correlation function [45] already without considering further cavity-QED effects:

$$g^{(2)}_{th}(0) = 2(1 - \frac{1}{N}). \tag{3.4}$$

Value between 1.9 and 1.95 are therefore expected if 20-40 QDs contribute to the emission in the II-VI cavity. Considering the QD density of this sample this number corresponds to roughly 8 − 17 % of the QDs inside the micropillar coupling to the the cavity mode. Considering the spectral overlap between the distribution of the QD emission energies and the cavity mode and possible cavity feeding effects, this value is reasonable. Interpretation of the results below threshold is more complicated for the III-V samples. For these structures there is no resolvable saturation of $g^{(2)}(0)$ below threshold. The quantum efficiency of the S-20 photocathode inside the streak camera is about one to two orders of magnitude worse for the wavelength range around 900 nm where the III-V samples operate compared to the 500 nm-range where the II-VI sample operates. Correspondingly background noise contributes more strongly and the necessary measurement duration becomes much longer already in the threshold region for the III-V samples. Under these circumstances, it becomes impossible to monitor the photon statistics below threshold. The highest measured values of $g^{(2)}(0)$ are about 1.7 for the low-Q sample and approximately 1.4 for the high-Q sample. A detailed theoretical analysis of the expected photon statistics

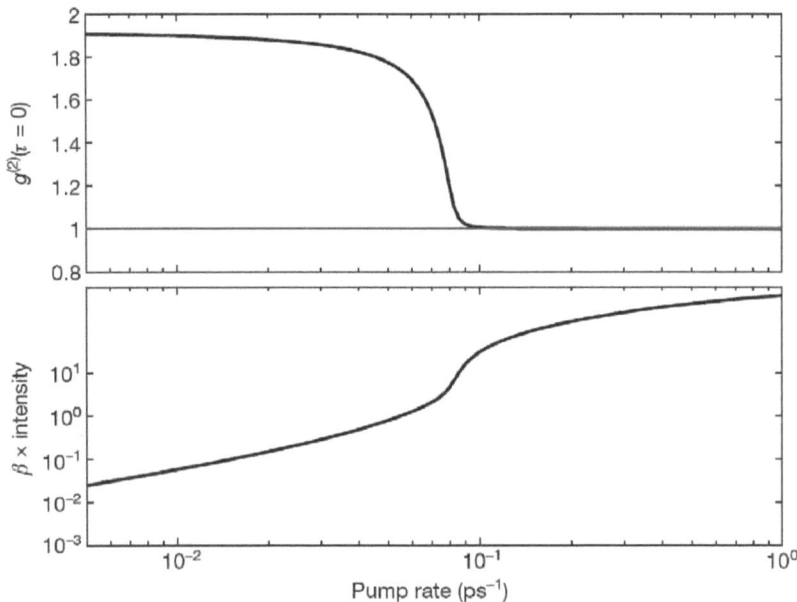

Figure 3.4: Calculated zero-delay correlation function and input-output curve for a low-Q quantum dot microcavity laser.

(see appendix A for details) for parameters as given by the samples used in our experiments which also takes cavity-QED effects into account shows that the measured values are not far from the saturation values [42]. The theoretical $g^{(2)}(0)$-curves are shown in the upper panels of figures 3.4 (assuming parameters of the low-Q cavity and 30 resonant QDs) and 3.5 (assuming parameters of the high-Q cavity and 15 resonant QDs), respectively. These curves saturate at values of ~ 1.9 for the low-Q sample and ~ 1.45 for the high-Q sample. Comparing the theoretical and experimental $g^{(2)}(0)$-curves at low excitation powers at comparable points in the IO-curve shows reasonable qualitative agreement between them. Inside the threshold region the II-VI and the low-Q III-V sample show the classically expected behavior: $g^{(2)}(0)$ undergoes a smooth transition towards a value of 1. The high-Q sample shows a different behavior. Here, $g^{(2)}(0)$ drops to values below unity around the lasing threshold, giving clear evidence for the emission of non-classical light. This is accompanied by an even smoother input - output curve. The calculations reproduce this peculiar dip. It can be traced back to the small number of emitters coupling to the cavity mode. This non-classical behavior vanishes with rising excitation intensity as the intracavity photon number increases and coherence starts to

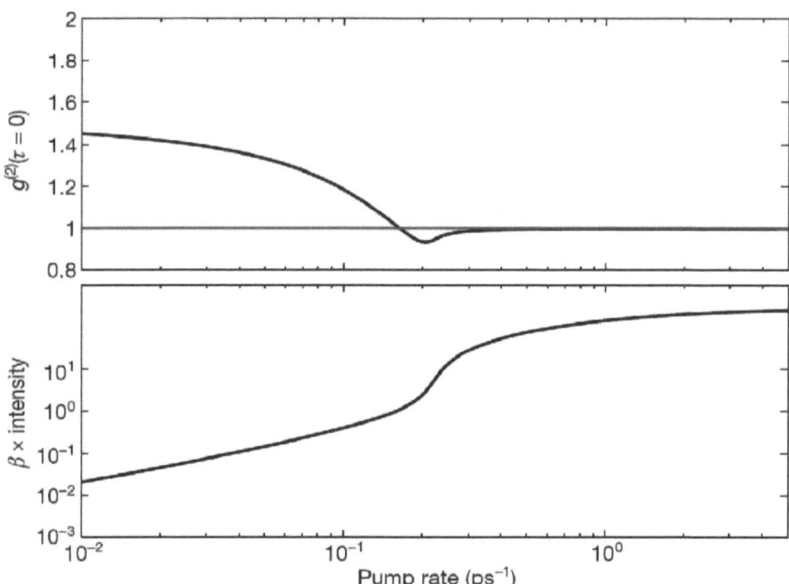

Figure 3.5: Calculated zero-delay correlation function and input-output curve for a high-Q quantum dot microcavity laser.

build up. Similar transitions from antibunching to bunching have been seen for atoms in a cavity [46] for a variable number of emitters. Further insight into the details of the emission process can be gained from time-resolved correlation measurements which will be the subject of the next section.

3.3 Time-resolved Correlation Measurements

For large τ, $g^{(2)}(\tau)$ must necessarily return to 1 as photons emitted with a large delay are statistically independent. For thermal light, straight application of classical coherence theory predicts a decay from the value at $\tau = 0$ towards a value of unity on a timescale of the order of the coherence time of the light. The dynamical evolution of the second-order correlations of the III-V cavities shown for selected excitation powers in figure 3.6, differ from this prediction. For the low-Q cavity (left panel) and excitation below the threshold region, $g^{(2)}(\tau)$ drops smoothly from a value slightly below two towards unity within the first picoseconds. However, for delays of around 40 ps, values below unity appear. For the high-Q cavity (right panel) and excitation in

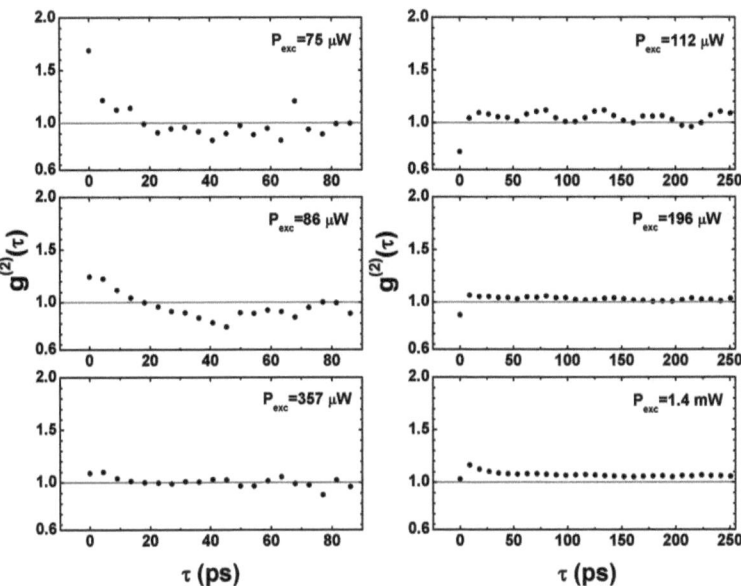

Figure 3.6: Temporal evolution of $g^{(2)}$ for selected pump powers. Results are shown for the low-Q (left panel) and high-Q (right panel) III-V cavities of figure 3.3. An unexpected outcome is the appearance of dynamical antibunching $g^{(2)}(\tau) < 1$ for low and intermediate excitation densities at finite τ for the low-Q cavity. The intensity correlation function does not merely drop to a value of unity, but takes on smaller values with subsequent oscillations. This is also particularly apparent for the high-Q cavity at $P_{exc} = 112\,\text{mW}$. In the threshold region the high-Q cavity shows antibunching caused by the small number of quantum dots coupling to the cavity.

the threshold region, pronounced long-lasting oscillations of $g^{(2)}(\tau)$ are visible. The oscillations become obviously damped with increasing excitation power. Above threshold, we find that $g^{(2)}(\tau) \approx 1$ in both the low-Q case and the high-Q case. The origin of these oscillations and the

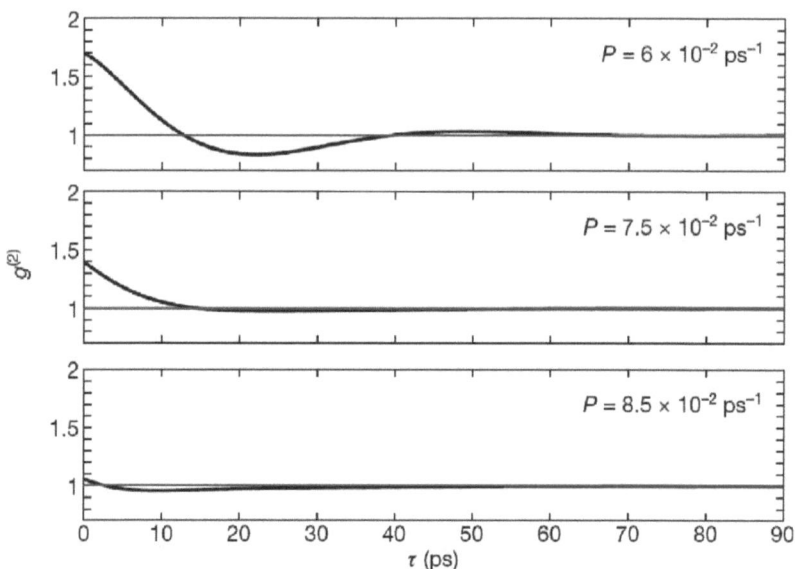

Figure 3.7: Calculated $g^{(2)}(\tau)$ for a low-Q quantum dot microcavity laser. Selected pump rates are used.

observed antibunching is not immediately clear. Application of a microscopic theoretical model (for details, see chapter A) gives deeper insight into the complex dynamics of the QD-cavity system. Figures 3.7 and 3.8 show examples of the numerical calculations. They are intended to demonstrate possible results for two different sets of parameters. The data are sensitive to the microscopic description of the carrier scattering that provides a common source for carrier redistribution and dephasing. These processes depend on, among other things, the electronic states both for the recombination processes and for where the carriers are pumped. To simplify this rather involved analysis, pumping at higher quantum-dot states was assumed. It was further assumed that 50 quantum dots in the low-Q micropillar and eight quantum dots in the high-Q micropillar are resonant with the optical mode. A spontaneous emission factor of $\beta = 0.1$, a total spontaneous-emission time (enhanced as a result of the Purcell effect) of $\tau_{sp} = 51.7$ ps (low Q) or 0.75 ps (high Q), and a quality factor of $Q = 40,000$ (low) or $80,000$ (high) were

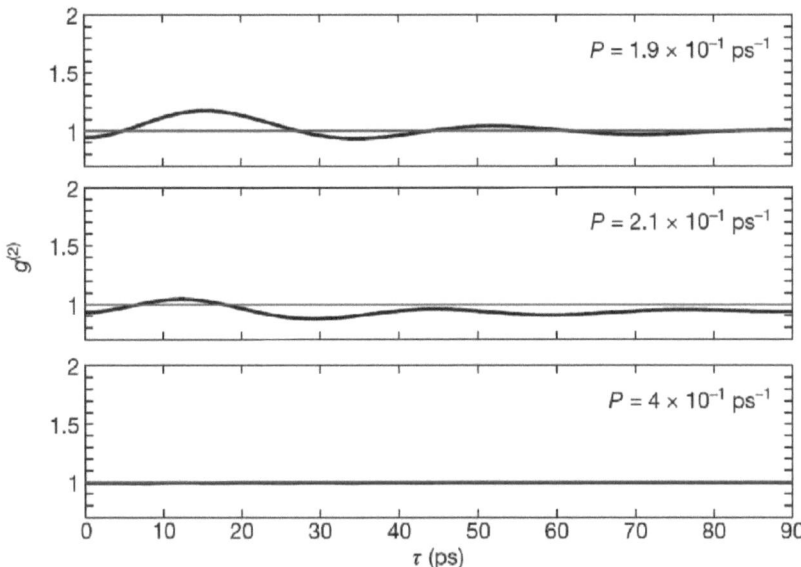

Figure 3.8: Calculated $g^{(2)}(\tau)$ for a high-Q quantum dot microcavity laser. Selected pump rates are used.

chosen for the calculations. Considering resonant pumping of carriers in the quantum-dot p shell, relaxation times from p shell to s shell of 0.5 ps for electrons and 0.25 ps for holes were used. The chosen examples reproduce the general trends of the experiments. The calculations of $g^{(2)}(\tau)$ for the low-Q case at low excitation intensities show a non-monotonic decay to unity from the initial value. For the high-Q parameter set oscillations in $g^{(2)}(\tau)$ are found as seen in the experimental data.

The observed oscillations in the photon correlations are a result of the dynamical coupling between photons and carriers and can be qualitatively understood as follows. For a microcavity operated at steady state in the spontaneous-emission regime, $g^{(2)}(\tau)$ decays from nearly two to one on the timescale of the coherence time. In the regime of dominating stimulated emission, $g^{(2)}(\tau)$ equals one independently of time delay. The oscillations are observed in the regime of transition from spontaneous to stimulated emission in a system that, under these conditions, contains only very few photons emitted by very few quantum dots. This is the transition regime of cavity QED lasers discussed in [2]. Unlike in the situation in the lasing regime, here the loss of a photon from the cavity represents a severe perturbation of the system, which strongly influences the coupled carrier-photon dynamics. Systems of emitters coupled to a cavity mode are known to exhibit different kinds of oscillations of the emission intensity. Relaxation oscillations can occur close to the threshold region when the laser is switched on or perturbed, and Rabi oscillations can occur in the regime in which the dissipation is small in comparison with the light-matter coupling strength. The dynamics of the correlation function $g^{(2)}(\tau)$ can be traced back to this behavior. Theory predicts that both kinds of oscillations can be triggered by photon emission events. In reaction to the perturbation, the system tries to re-establish equilibrium, and, in doing so, undergoes quantum oscillations. In both cases, the origin of these oscillations is the feedback due to the cavity, which can lead to out-of-phase oscillations of photon number and lasing medium. These oscillations become damped as the pump rate increases and a regime of stimulated emission is reached in which the photon number is high enough that single photon losses no longer affect the system considerably. In the case discussed before, the perturbation of the few-emitter system can become so prominent that, for example, the subsequent emission of a photon is suppressed. This leads to the dynamic antibunching both for zero delay and for times after enhanced photon pair emission during the oscillations. These oscillations potentially also appear in intensity measurements. However, as the moment of photon emission is stochastic, any time averaging blurs the oscillations. Nevertheless, the oscillations also carry over to the correlation functions similar to $g^{(2)}(\tau)$. There they survive the averaging, as although the moment of arrival of the first photon is still stochastic, a second photon is picked whose delay, τ, relative to the first is fixed for all detected photon pairs. For increasing Q, the cavity feedback is enhanced, causing the quantum fluctuations to become more pronounced as reflected by the oscillations of $g^{(2)}(\tau)$.

The experimental and theoretical results shown so far show reasonable qualitative agreement although measurements using pulsed excitation were compared with the results of steady state calculations. It is also worthwhile to study a regime where differences between pulsed and steady state operation occur, namely the dynamics in the build-up and break-down of coherence during a pulse. The starting point of such a study is again the time-resolved recording of the individual photon emission events in the output pulse which provides a complete mapping of the second-order correlation function of the emitted light in t and τ. Instead of determining the τ-dependence by calculating the intensity-weighted average of $g^{(2)}(t,\tau)$ over a time interval t during the emission pulse, as done in the earlier discussions, now the t-dependence at a vanishing photon pair delay time of $\tau = 0$ is the central quantity to be studied. While the former treatment gives a good characterization of the photon statistics and coherence time of the emitted pulse as a whole, the latter is a good measure of the time-resolved second-order coherence properties and therefore also the build-up and break-down of coherence during a pulse. For these measurements a micropillar with 6 μm diameter was used. All other characteristics of this pillar are comparable to the low-Q III-V micropillar discussed before. The transition region to lasing was identified by measuring the input-output curve shown in Fig. 3.9. A nonlinear behavior between excitation powers of 60 and 150 μW marks this region. At

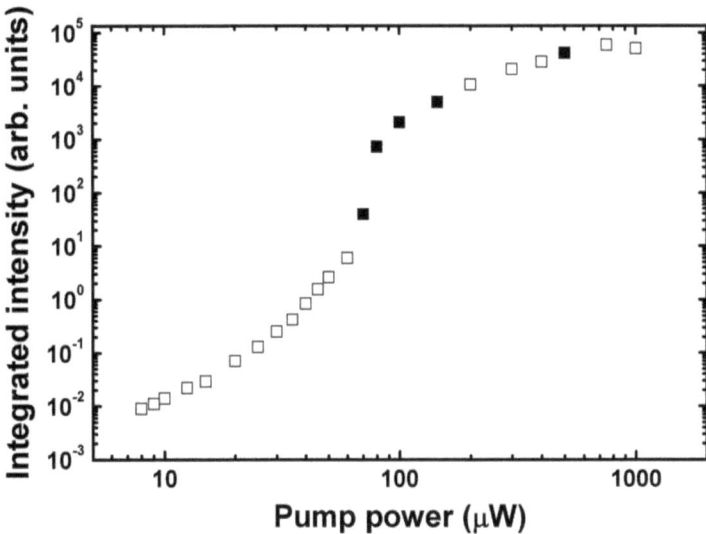

Figure 3.9: Integrated intensity of the 6 μm pillars fundamental mode under nonresonant pulsed excitation. Above threshold, saturation effects become apparent. Filled squares mark the data sets shown in figure 3.10.

lower excitation powers, $g^{(2)}(t,0)$ takes on the expected static value of 2 without showing any dynamics. More interesting are the excitation powers (filled squares in figure 3.9) in and above the threshold region for which the intensity-dependent second-order coherence properties of the emission were determined. Figure 3.10 shows $g^{(2)}(t,0)$ for these excitation powers alongside the temporal emission-intensity profiles. The light exhibits thermal behavior at the very beginning and at the very end of the emission pulse. After the generation of carriers in the barrier states by the pump pulse, these carriers rapidly relax into the QD states [47]. As long as a small number of carriers is present in the QD states, spontaneous recombination processes determine

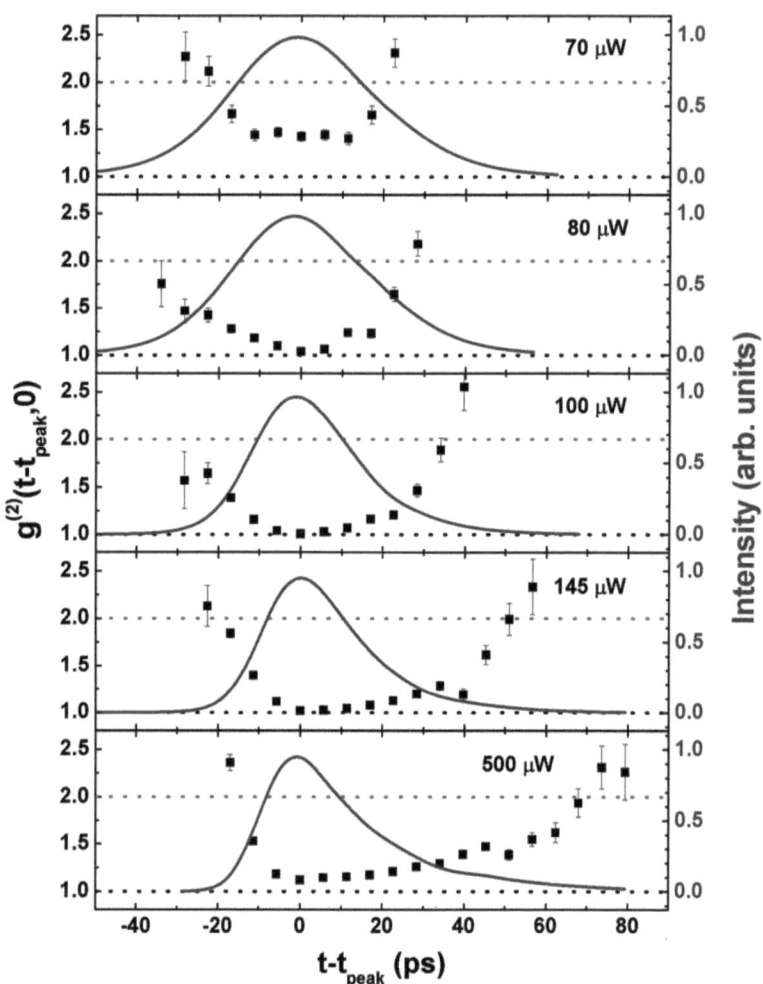

Figure 3.10: Time evolution of the second-order photon correlation function $g^{(2)}(t,0)$ (symbols) compared to the normalized output intensity (solid lines) for the micropillar fundamental mode. Black and red dotted lines denote the limiting cases of $g^{(2)}(t,0)$ for coherent and thermal light, respectively. The power density for pulsed excitation increases from top to bottom. t_{peak} corresponds to the maximum of the emission intensity for each pump power.

the output. When the population becomes sufficiently strong, the system is driven into the regime of coherent emission, characterized by a decrease in the second-order correlation function toward the value of 1, and by a faster decay of the emission intensity due to the stimulated processes, visible in an apparent temporal narrowing of the emission peak. The decrease toward $g^{(2)}(t,0) = 1$ becomes more pronounced for higher excitation densities, finally leading to a broadening of the dip, as more carriers are excited in the system and stimulated emission can be maintained for a longer time. While the transition from thermal to coherent emission in the beginning of the pulsed emission can take 40 ps or even more in the threshold region, it happens on a time scale on the order of 10-15 ps far above threshold.

As another important finding, it should also be noted that knowledge of $g^{(2)}$ allows us to determine how large the relative amounts of coherent and thermal emission are for any given time within the emission pulse. This information is not accessible via output intensity measurements alone. One can consider partially coherent light as a superposition of a thermal and a coherent mode, which contribute to $g^{(2)}$ according to [48]

$$g^{(2)}(t,0) = (1 + \frac{1}{\gamma} - \frac{1}{2\gamma^2} + \frac{e^{-2\gamma}}{2\gamma^2})R_t^2 + g_c^{(2)}(t,0)R_c^2 + 2R_t R_c (1 + \frac{2e^{-\gamma}}{\gamma^2} + \frac{2}{\gamma} - \frac{2}{\gamma^2}), \quad (3.5)$$

where $\gamma = \Gamma T$ with $\hbar\Gamma \approx 63\,\mu\text{eV}$ is the thermal mode half width at half maximum, $T = 5\,\text{ps}$ is the sampling time and R_c and R_t are the relative fractions of coherent and thermal emission, respectively. As can be seen in figure 3.11, the thermal and coherent fractions depend nonlinearly on $g^{(2)}$ showing that already small amounts of thermal emission can cause significant deviations from coherent emission. The finite sampling time causes a slight underestimation of $g^{(2)}$ for large thermal fractions. In our case, the minimum $g^{(2)}$ for 70 μW excitation power is about 1.5, for which about 70 % of the emission is coherent. When $g^{(2)}$ drops on the other hand below 1.2, more than 90 % of the emitted light is coherent.

It should be noted that in the time evolution of the second-order coherence $g^{(2)}(t,0)$, the position of the minimum coincides with the peak of the mean photon number of the emission pulse. Also the time dependence of the leading and trailing edges of the emission pulse is mirrored in the dynamics of $g^{(2)}(t,0)$. This suggests that the interplay of stimulated emission and output coupling determines not only the emission intensity but also the coherence properties. This behavior appears surprising as the dynamical evolution of the mean photon number $\langle \hat{b}^\dagger(t)\hat{b}(t)\rangle$ and of the two-photon coincidences $\langle \hat{b}^\dagger(t)\hat{b}^\dagger(t)\hat{b}(t)\hat{b}(t)\rangle$ obey distinctly different equations of motion in which different types of carrier-photon correlations enter, even though both expectation values can be traced back to the time dependence of the photon probability distribution.

At the very early and late parts of the emission pulse, where low values of the emitted intensity

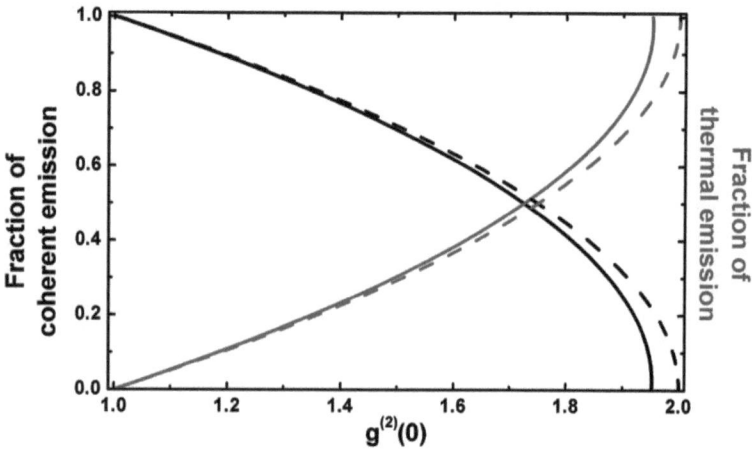

Figure 3.11: Relative fractions of coherent and thermal emission at a fixed $g^{(2)}(t,0)$ as given by a two-mode model (solid lines) compared to the ideal case for infinitely small sampling time (dashed lines). Small deviations occur at high thermal fractions.

are present, an overshoot of $g^{(2)}(t,0)$ beyond the thermal value of 2 is obtained in the experiment. This behavior is an artifact of pulse jitter caused by electronic noise as discussed in detail in section 2.2.3 and depicted schematically in figure 2.17. However, the emitted pulses are longer and not Gaussian, so that a qualitative discussion of this overshoot under these conditions is necessary. The detected mean photon pair and photon count rates at a screen position corresponding to time t are in fact a mixture of all photon pair and photon count rates weighted with a narrow Gaussian distribution centered at t. To determine whether this jitter has a significant effect on the recorded photon statistics, it is necessary to compare the time scale on which the jitter occurs to the time scale of the pulse dynamics. If the pulse dynamics are comparable to the jitter time scale or even faster, the momentary intensity at a certain position on the screen will vary strongly from picture to picture and the measured correlation function will depend on these fluctuations instead of the intrinsic fluctuations of the light field. For visualization of this effect, one can consider a coherent pulse with varying intensity and a simplified jitter model, which leads to well defined shift Δt with a probability p and causes no shift at all other times with the probability $q = 1 - p$. The measured intensity correlation at a position on the screen corresponding to time t will now only depend on p and the mean photon number ratio $r = \frac{n(t+\Delta t)}{n(t)}$ of the times connected by the jitter,

$$g^{(2)}(t,0) = \frac{qn(t)^2 + p[rn(t)]^2}{[qn(t) + prn(t)]^2} = \frac{q+pr^2}{(q+pr)^2}. \tag{3.6}$$

This function is depicted in figure 3.12 for rare events (dotted line) and common events (solid line). The detrimental effect of the common jitter shown there is negligible for our measurement

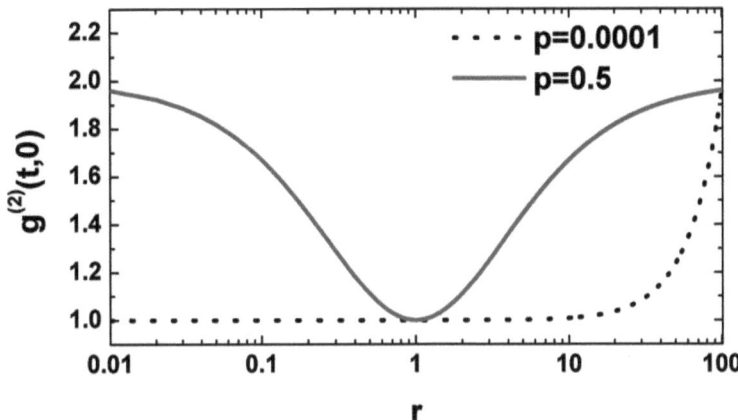

Figure 3.12: Effect of jitter on the measured $g^{(2)}(t,0)$ for a coherent light pulse depending on the ratio r of the mean photon count rates at the pulse positions connected by the jitter and the relative frequency p of jitter occurrence. While frequent jitter (solid line) has the same effects for pulse positions with high and low intensity, rare events (dotted line) only affect regions with low mean photon count rate (large r).

because common jitter happens on a time scale of about 1.5 ps and the mean photon count rates do not change significantly in this range. Even at the steepest positions of the pulse slope, the intensity variation does not exceed 7 % within 1.5 ps (compare figure 3.10). Accordingly, only the region between $r = 0.93$ and $r = 1.07$, where the red line does not show significant deviations from the expected value of 1, contributes for frequent jitter. For rare events on the other hand, there is no effect for small r but there are significant deviations for $r \geq 20$. As Gaussian jitter is unbounded, there are indeed rare jitter events where r exceeds 20 for regions with small mean photon count rate. These rare events cause the overshoot of $g^{(2)}(t,0)$ seen far from the emission peak in figure 3.10. In these regions, the increased pair detection rate due to rare jitter events is larger than the intrinsic photon pair count rate determined by the mean intensity at this position. This consideration shows that special care must be exercised when the statistics of weak emission signals are studied.

The same microscopic theory applied to model the relaxation oscillations and the occurrence of antibunching discussed before, has also been used to model the dynamics of coherence build-up and break-down. As the details of carrier relaxation and capture processes are not crucial for the photon-correlation dynamics, incoherent generation of carriers in excited QD states with a time-dependent rate determined by an effective pump pulse is assumed. The model includes subsequent carrier relaxation into the lowest QD states that are coupled to the fundamental cavity mode. Carrier scattering and dephasing rates are obtained from independent many-body calculations. A pulse duration of 68 ps provides a reasonable estimate for the population dynamics of the excited QD states to obtain an evolution of the output intensity as observed in the experiment. This value is in reasonable accord with recent studies on similar QD structures, which give luminescence rise times on the order of 50 ps [49]. The calculated time-resolved intensity and second-order correlation function $g^{(2)}(t, \tau = 0)$ are shown in figure 3.13. For increasing values of the time-integrated pump rate from $P = 3$ to $P = 7$, a temporal narrowing of the output pulse due to the increasing stimulated-emission contribution accompanied by a growing dip in the photon-correlation function approaching unity at the intensity maximum is found. With increasing pump rate, inversion of the carrier states is reached earlier and persists longer. For the highest considered pump rate $P = 10$, saturation effects due to Pauli blocking start to reduce the peak intensity of the output pulse. As a consequence, these saturation effects also lead to the appearance of a longer pulse duration.

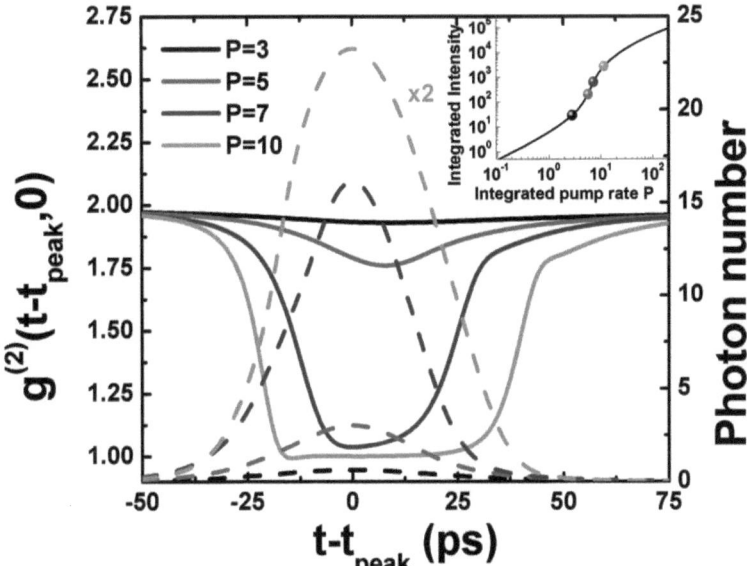

Figure 3.13: Calculated input-output curve (inset) and time evolution of the mean photon number (dashed lines) and the second-order correlation function (solid lines) for selected values of the time-integrated pump rate indicated by the color-coded dots on the input-output curve. All curves are calculated for a pump-pulse width of 68 ps. The following parameters were used: number of resonant QDs: 140, cavity loss rate 2κ: 0.09 ps^{-1}, spontaneous lifetime: 4.3 ps, scattering rates into (out of) the lasing transition: 11.7 ps^{-1} (23.3 ps^{-1}), light-matter coupling constant g: 0.22 ps^{-1}, loss rate to nonlasing modes: Γ_{nl}=0.22 ps^{-1}, and constant dephasing: Γ=8.5 ps^{-1}. Comparison with extended calculations including an inhomogeneously broadened ensemble gives very similar results with the present calculations using a homogeneous ensemble subject to considerable dephasing.

Chapter 4

Quantum Well Diodes and VCSELs

In contrast to VCSELs, quantum-well polaritonic diodes operate in the strong coupling regime. Although often only polaritonic devices using electrical injection [50, 51] are considered polaritonic diodes, in the following there will be no distinction between polaritonic devices using electrical and nonresonant optical injection and both will be termed polaritonic diodes. Both kinds of diodes rely solely on spontaneous relaxation mechanisms to guide the excited carriers to the emission modes. Usually, the lower polariton branch emission is the quantity of interest. Figure 4.1 shows a schematic picture of the relaxation mechanisms [52]. The excitation is provided by external pumping and creates a thermal reservoir of free electrons and holes. Polaritons are formed from the thermal reservoir via acoustic- or optical-phonon emission. Inside the polariton branches acoustic-phonon scattering and radiative recombination are the most important processes. Other scattering processes, in particular exciton-carrier scattering, exciton dissociation, exciton-exciton and polariton-polariton scattering, can be neglected at low lattice temperatures and carrier densities. As the $k_{\parallel} = 0$ lower polariton state shows the lowest energy, one might imagine this state to act as a polariton trap and expect most emission to come from this state. Experimental results show a different situation with the major part of the emission stemming from the so-called bottleneck region marked in figure 4.1 for negative detunings [53]. These experimental findings can be explained by the taking the acoustic-phonon scattering rates for polaritons into account. The maximum energy transferred by acoustic-phonon scattering is on the order of $1\,\text{meV}$ [54]. In the bottleneck region the energy changes very fast with $k_{\parallel}$, resulting in a small density of states and also in a reduced acoustic-phonon scattering rate. Additionally also the polariton lifetime decreases strongly in this region due to the large photonic content as shown in figure 1.6 causing the multiple phonon scattering processes needed to realize relaxation to the ground state to be very unlikely. This relaxation bottleneck

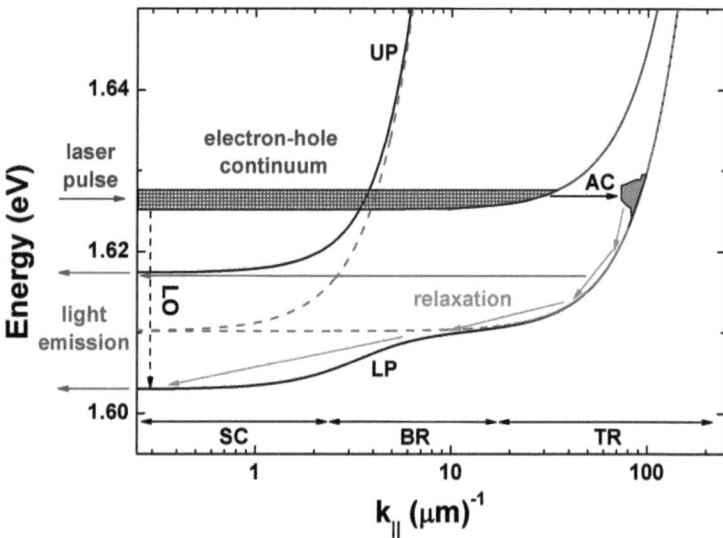

Figure 4.1: Schematic representation of the polariton dispersion and the possible polariton formation and relaxation processes by means of optical (LO) and acoustic (AC) phonon emission on a logarithmic scale. Dashed lines give the bare cavity and exciton dispersions. LP and UP mark the lower and upper polariton dispersion, respectively. The plaid region marks the thermal free carrier reservoir. SC, BR and TR mark the strong-coupling, bottleneck and thermal regions.

becomes more pronounced for larger negative detunings because the density of states in the low-$k_{\parallel}$ region decreases strongly with increased photonic content of the lower polariton. Stimulated polariton-polariton scattering processes provide a way towards more efficient relaxation towards the ground state as the maximum amount of energy transferred in a single scattering process is larger. However, such processes become efficient only at larger polariton densities than are realizable in polaritonic diodes. For high polariton densities the Coulomb interaction between the excitonic fractions of the polaritons increases and finally leads to a bleaching of the exciton oscillator strength [55] which in turn leads to a reduction of the Rabi splitting and finally breaks the strong coupling regime. The polaritonic diode then transits into the weak coupling regime and turns into a VCSEL. For polaritonic diodes the strong coupling regime is bleached before the polariton density becomes high enough to allow stimulated polariton scattering processes under nonresonant excitation conditions.

4.1 Correlation measurements on QW VCSELs

The QW VCSEL sample consists of a GaAs/ AlGaAs microcavity grown by molecular beam epitaxy. It contains one 10-nm-wide quantum well placed in the electric field antinode of a slightly wedged λ cavity especially designed to avoid charge accumulation in the quantum well [56]. The sample displays a vacuum Rabi splitting of 3.9 meV. The polariton dispersion for different excitation densities (figure 4.2) shows an apparent bleaching of the strong coupling regime with increasing excitation power. Additionally, the LP ground state was found to be only weakly populated far below the lasing threshold. Therefore, polariton-polariton scattering is also weak in this regime. The far-field emission of the LP branch was investigated at a negative detuning of -2 meV. The Fourier plane of the emission was either imaged onto the entrance slit of a monochromator for measuring the dispersion or onto the entrance slit of a streak camera for photon counting measurements as explained in chapter 2. Photons, which are emitted at an angle of θ, directly correspond to polaritons with energy E and in-plane wave vector of $k_{\parallel} = \frac{E}{\hbar c}\sin\theta$. Thus, in the first case, the entrance slit of the monochromator selects a narrow stripe with $k_{x,\parallel} = 0$. In the second case, only the $k_{\parallel} = 0$ state of the LP branch is selected with an angular resolution of $\sim 1°$ by using a pinhole. Additionally, an interference filter with a 1-nm width is used to ensure that only a single mode contributes to the signal. With increasing excitation density, the filter is tuned so that the central transmission wavelength follows the blue shift of the polariton dispersion as shown in figure 4.3. As can be seen, there is a smooth and continuous blue shift starting at the onset of the nonlinear region in the input-output curve. This is a clear sign of increasing interactions between particles and bleaching of exciton oscillator strength. The measured time-averaged normalized intensity correlation functions $g^{(n)}(\tau = 0)$ up to the fourth order (figure 4.4) show that, for high excitation densities, all orders approach the expected value of 1, denoting conventional photon lasing. With decreasing excitation density, a smooth transition toward the thermal regime occurs, which is accompanied by photon bunching. At an excitation power of $\sim 1.5\,\mathrm{mW}$, the bunching effect saturates at values of approximately 2 and 6, which are the expected values of n factorial for the second and third orders of $g^{(n)}(\tau = 0)$. The fourth order also shows an increase of the joint detections, but the number of detected four-photon combinations is too small at low excitation densities to give statistically significant results in the thermal light regime. The results for different orders of $g^{(n)}(\tau = 0)$ at the same excitation power are derived from the same data set. To assure that

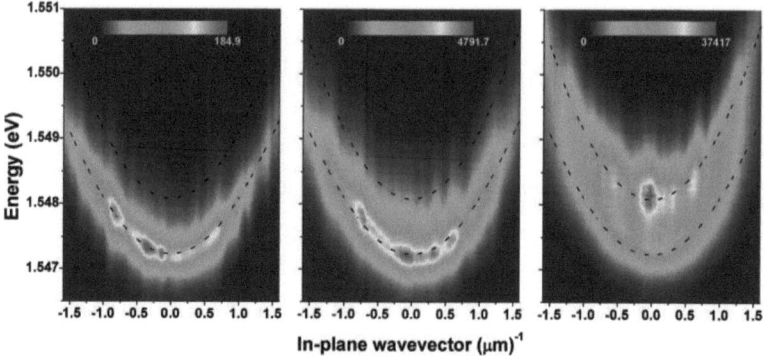

Figure 4.2: Momentum distribution of the polaritons as measured by angle-resolved photoluminescence for three different excitation densities: (left panel) 50 mW (far below the lasing threshold), (middle panel) 1.5 mW (at the lasing threshold), and (right panel) 10 mW (above threshold). The false color scale is linear. The black dashed lines indicate the dispersion of the LP and the bare cavity mode.

only single-mode thermal emission from the $k_\parallel = 0$ state was measured, the collection angle was also increased by opening the pinhole. By doing so, the number of modes contributing to the signal increases and, therefore, the regime of indistinguishable photons is left. As can be seen in the right inset of figure 4.4, photon bunching is only present at collection angles below 1.5°, suggesting that we are indeed operating in the single-mode thermal regime. To further ensure that the experimental results are the result of photon bunching and not just a consequence of some dominating noise source when the signal gets weaker, we also studied the blue shift of the LP and the input-output curve of the microcavity, as shown in figure 4.3. The onset of the decrease of $g^{(n)}(\tau - 0)$ coincides with the beginning of the LP blue shift and the onset of a nonlinear increase in the input-output curve. This shows that the system leaves the strong coupling regime and starts to lase. At high excitation powers, well defined lasing at the bare cavity mode builds up as expected. It is obvious that the thresholds,where the $g^{(n)}(\tau = 0)$ begin to decrease toward a value of 1, do not occur at the same excitation density. This shift can be explained in terms of the low photon numbers inside the cavity at the lasing threshold. Stimulated emission sets in at a mean photon number p of the order of unity inside the mode of interest, but in the threshold region, there is still a superposition of thermal and stimulated

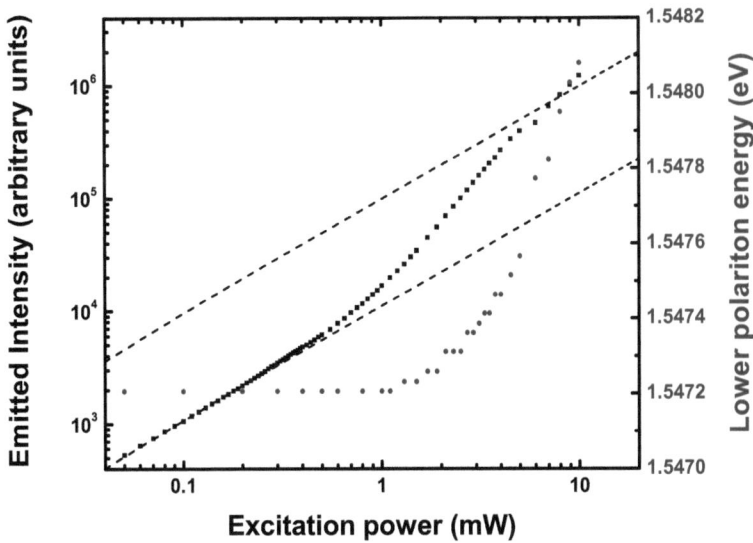

Figure 4.3: Squares indicate integrated intensity of the lasing mode measured at normal incidence as a function of the nonresonant excitation power at a detuning of -2 meV. Circles represent emission energy around normal incidence as a function of the excitation energy. Dashed lines indicate the linear dependence of the emitted intensity on the excitation power below (lower curve) and above (upper curve) the lasing threshold.

emission present. Because of the stronger photon number fluctuations in chaotic fields, their contribution to n-photon combinations will still be substantial, whereas p is smaller than n.

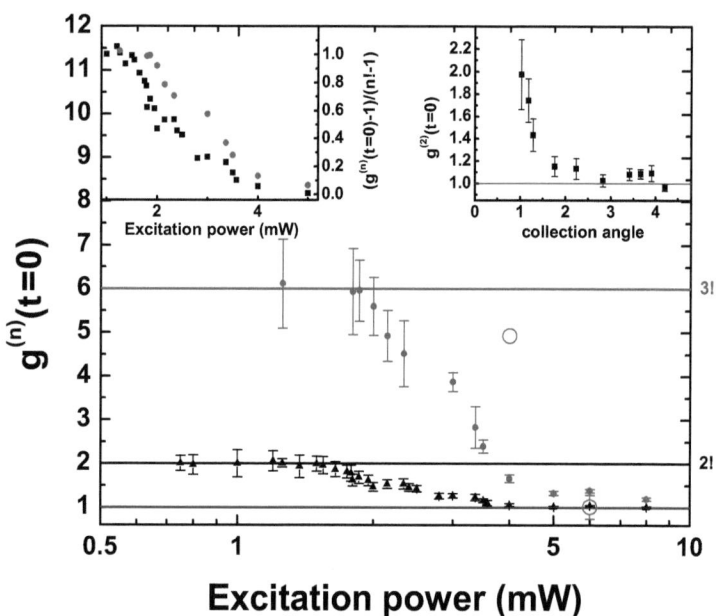

Figure 4.4: Second (triangles), third (spheres) and fourth (open circles) order intensity correlation function versus excitation power. (Left inset) Close-up of the second- (squares) and third-order (spheres) intensity correlation on a normalized linear scale. A value of 0 corresponds to $g^{(n)}(0) = 1$ and a value of 1 corresponds to $g^{(n)}(0) - 1$. (Right inset) Second order intensity correlation function in the thermal regime for several total collection angles. Error bars indicate the variation of the correlations, resulting from the standard deviations of the detected photon numbers used for calculating $g^{(2)}$ and $g^{(3)}$.

Chapter 5

Polaritonic Condensates

5.1 Nonequilibrium condensation

In the low-density limit cavity polaritons approximately behave like bosons. Thus, they can in principle undergo Bose-Einstein condensation (BEC) when their de-Broglie wavelength becomes comparable to their average separation. Their dual light-matter nature makes this approach very promising. This mixture results in a small mass and allows for very high critical temperatures up to room temperature. Also it is easily accessible experimentally because the emitted photons are part of the polariton wavefunction [57] and the properties of the emitted light directly reflect the properties of the polaritons. The drawback of the added photonic content is the short lifetime of the polaritons. It is typically on the order of 2 - 10 ps, which makes it difficult to establish thermal equilibrium. In fact, full thermalization with the host lattice has not been realized for polariton condensates. It is, however, possible to create a polariton gas in self-equilibrium if polariton-polariton scattering processes are fast enough.

5.2 Strategies to reach degeneracy

After the first proposal of polariton lasers or polariton BEC acting as a laser without inversion [58] huge efforts were devoted to the development of suitable structures to realize this scenario. Basically the experimental structures are still similar to the ones used for realization of QW diodes in chapter 4. However, the main difficulty lies in reaching a quantum degenerate ground state and bosonic final state stimulation before significant exciton oscillator strength bleaching sets in. There are several strategies to overcome this problem and achieve large polariton densities and efficient stimulated polariton-polariton scattering. They can be divided into two main categories: approaches aiming at realizing a larger Rabi splitting by a certain choice of material or design of the cavity and approaches utilizing a more efficient excitation scheme. The description of microcavities given in section 1.2.1 focused on GaAs based approaches as it is a material allowing for very precise sample growth with a small defect density and most samples investigated in this thesis are GaAs based. However, the achievable $2\hbar\Omega_R$ for a single quantum well is on the order of $3 - 4\,\text{meV}$ only. Correspondingly there have been considerable efforts to realize larger $\hbar\Omega_R$ in the strong coupling regime using materials with larger exciton oscillator strength like CdTe ($2\hbar\Omega_R$ up to $25\,\text{meV}$) [59], GaN ($2\hbar\Omega_R$ exceeding $50\,\text{meV}$) [60] and ZnO ($2\hbar\Omega_R$ up to $50\,\text{meV}$) [61]. Another strategy also used for GaAs systems lies in increasing the number N_{QW} of quantum wells embedded in the cavity [62]. As Ω_R increases with the square root of the exciton oscillator strength, adding more of them corresponds to an effective coherent addition of the single quantum well oscillator strengths while simultaneously the polariton density per quantum well at fixed total polariton number scales as N_{QW}^{-1}. A common strategy is to position three stacks of 4 quantum wells each at the central antinode of the cavity and the first antinode in each DBR mirror. By doing so Rabi splittings as large as $15\,\text{meV}$ have been realized for GaAs systems, allowing to reach stimulated scattering into the LP ground state.

There are basically four different excitation schemes using optical excitation with varying efficiency. However, not all of them are suitable for each kind of experiment. The easiest method is direct resonant excitation of the $k_{||} = 0$ condensed state. While being very efficient, this excitation geometry is not suitable for proof-of-principle experiments aimed at demonstrating features like spontaneous build-up of coherence as the coherence properties of the ground state could have been inherited directly from the resonant optical pumping process. The same problem arises for resonant pumping at the so-called magic angle. The magic angle state $k_{||,M}$ is the point of the LP dispersion from which a resonant polariton-polariton scattering process to the states $k_{||} = 0$ and $2k_{||,M}$ is possible. In this process both total wavevector and energy are conserved. However, it is a resonant scattering process and it is still possible that the coherence of the pump beam is directly carried over to the condensed state. Therefore, only two pumping schemes remain to demonstrate spontaneous build-up of coherence: nonresonant pumping and resonant pumping of the LP under high angles where direct resonant scattering to the ground state is forbidden. In both cases several polariton-polariton or polariton-phonon scattering processes are necessary before the ground state is reached. During this processes the initial coherence is lost. Both excitation schemes have different pros and cons. Nonresonant excitation is very efficient. Here the initial excitation creates electrons and holes. For those, the density of states is significantly higher than for polaritons in a certain state. Accordingly threshold densities are at least a factor of 10 smaller compared to resonant excitation of polaritons with large $k_\parallel$. On the other hand, nonresonant excitation also creates a large number of background carriers interacting with the polariton gas. Therefore the exact features of the polariton BEC will also depend strongly on the density and spatial distribution of the residual carriers. Resonant excitation of polaritons with large $k_\parallel$ is less efficient and results in large threshold excitation densities, but also offers well defined experimental conditions. Polaritons can be created directly with a desired polarization and there is no excitation of other carriers interacting with them. In this chapter both excitation schemes will be applied.

5.2.1 Definitions and signatures of BEC

The definition and experimental identification of BEC is nontrivial. Over the course of the last decades several possible definitions of BEC have been proposed which are connected with different experimental signatures. This section will roughly follow the account given in [63]. The first prediction of BEC goes back to Bose [64] and Einstein [65]. Considering N noninteracting bosons at a temperature T in a volume R^d consisting of the system size R and its dimensionality d, the energetic distribution of these bosons will be given by the following Bose-Einstein distribution:

$$f_B(\vec{k},T,\mu) = \frac{1}{\exp\left(\frac{E(\vec{k})-\mu}{k_B T}\right) - 1}. \quad (5.1)$$

Here $E(\vec{k})$ gives the bosonic dispersion with the lowest value of $E(\vec{k})$ set to 0, $\vec{k}$ is the particle wavevector also of dimensionality d and μ describes the chemical potential. $-\mu$ is the energy needed to add another particle to the system. For bosons, this definition makes sense only for nonpositive values of the chemical potential. Its exact value is determined by the normalization condition for a given total particle number N:

$$N(T,\mu) = \sum_{\vec{k}} f_B(\vec{k},T,\mu). \quad (5.2)$$

As the quantity of interest is the particle density of the ground state, it is convenient to divide between the ground state and all other states:

$$N(T,\mu) = \frac{1}{\exp\left(-\frac{\mu}{k_B T}\right) - 1} + \sum_{\vec{k}, \vec{k} \neq 0} f_B(\vec{k},T,\mu). \quad (5.3)$$

The particle density can be evaluated by taking the thermodynamic limit and converting the sum to an integral over reciprocal space:

$$n(T,\mu) = \lim_{R \to \infty} \frac{N(T,\mu)}{R^d} = n_0 + \frac{1}{(2\pi)^d} \int_0^\infty f_B(\vec{k},T,\mu) d\vec{k}. \quad (5.4)$$

Here the ground state particle density n_0 is given by:

$$n_0(T,\mu) = \lim_{R\to\infty} \frac{1}{R^d} \frac{1}{\exp\left(-\frac{\mu}{k_B T}\right) - 1}. \tag{5.5}$$

Well away from zero chemical potential, the ground state particle density vanishes, while the integral on the right-hand-side of equation 5.4 increases with μ. Accordingly, increasing the particle density n in the system will also cause μ to increase. As only nonpositive values of μ give sensible results, the maximum particle density n_c that can be accomodated following the Bose distribution function is given by:

$$n_c(T) = \lim_{\mu\to 0} \frac{1}{(2\pi)^d} \int_0^\infty f_B(\vec{k},T) d\vec{k}. \tag{5.6}$$

It seems that after reaching n_c no additional particles can be added. Einstein proposed that further increase of the particle density causes the added particles to collapse into the ground state, which has a density given by:

$$n_0(T) = n(T) - n_C(T). \tag{5.7}$$

This is a phase transition which main characteristic is the massive accumulation of particles in the ground state. The order parameter is the chemical potential, which vanishes at the transition. From an experimental point of view, however, this macroscopic occupation of the ground state is a rather qualitative sign of BEC as the exact onset of what should be considered as a large enough population of the ground state is not necessarily defined well under all circumstances. A more modern definition of BEC working also for interacting bosons which are not necessarily in equilibrium makes use of the single-particle density matrix [66]. Any pure state s of a system of N bosons at positions $\vec{r}_i$ (i=1..N) can be written in the form:

$$\psi_N^s(t) = \psi_s(\vec{r}_1, \ldots, \vec{r}_N, t) \tag{5.8}$$

The most general state of the system is given by a mixture of different of those normalized and mutually orthogonal pure states s with weights p_s. The single-particle density matrix $\rho_1(\vec{r}, \vec{r}', t)$ is given by:

$$\begin{aligned}\rho_1(\vec{r},\vec{r}',t) &= N \sum_s p_s \int dr_2 \ldots dr_N \psi_s^*(\vec{r},\vec{r}_2,\ldots,\vec{r}_N,t) \psi_s^*(\vec{r}',\vec{r}_2,\ldots,\vec{r}_N,t) \\ &= \langle \hat{\psi}^\dagger(\vec{r},t)\hat{\psi}(\vec{r}',t)\rangle \, (\vec{r}=\vec{r}_1, \vec{r}'=\vec{r}_1')\end{aligned} \tag{5.9}$$

in terms of bosonic field operators. The choice not to integrate over $\vec{r}_1$ is arbitrary. The single-particle density matrix is equivalent for each choice of $\vec{r}_j$. Basically the single-particle density matrix gives the product of the probability amplitudes to find a certain particle at position $\vec{r}$ and $\vec{r}'$, averaged over the behavior of all other particles. ρ_1 can be considered as a matrix with respect to $\vec{r}$. This matrix is Hermitian and can be diagonalized:

$$\rho_1(\vec{r}, \vec{r}', t) = \sum_i n_i(t) \chi_i^*(\vec{r}, t) \chi_i(\vec{r}', t). \qquad (5.10)$$

The eigenfunctions $\chi_i(\vec{r}, t)$ form a complete, orthogonal set for each time t. The eigenvalues of ρ_1 allow to characterize the system. Roughly speaking, the $n_i(t)$ are either of order unity or of order N, which is equivalent to the limiting value of (n_I/N) being either a constant or 0 in the thermodynamic limit. If there is one eigenvalue of order N the system exhibits BEC. This approach also allows to take fragmented BECs with more than one eigenvalue on the order of N into account, which are difficult to handle using other definitions of BEC.

Another possibility to classify BEC is the order parameter

$$\psi(\vec{r}, t) = \sqrt{N_0(t)} \chi_0(\vec{r}, t), \qquad (5.11)$$

where $\chi_0 = |\chi_0| \exp(i\phi(\vec{r}, t))$ is the eigenfunction of a single condensed state with particle number N_0. This macroscopic phase of the condensed state leads to other possibilities to spot BEC properties. The condensate density and the current carried by the condensed particles are given by:

$$\rho_c(\vec{r}, t) = N_0(t) |\chi_0(\vec{r}, t)|^2 \qquad (5.12a)$$
$$\vec{j}_c(\vec{r}, t) = N_0(t) \left(-\frac{i\hbar}{2m} \chi_0^*(\vec{r}, t) \nabla \chi_0(\vec{r}, t) + c.c. \right)$$
$$= N_0(t) |\chi_0(\vec{r}, t)|^2 \frac{\hbar}{m} \nabla \phi(\vec{r}, t), \qquad (5.12b)$$

respectively. The ratio $\vec{j}_c(\vec{r}, t)/\rho_c(\vec{r}, t)$ is termed the superfluid velocity $\vec{v}_s(\vec{r}, t)$. It does not depend on the magnitude of the order parameter:

$$\vec{v}_s(\vec{r}, t) = \frac{\hbar}{m} \nabla \phi(\vec{r}, t). \qquad (5.13)$$

This is an important quantity when discussing superfluidity. There are two important conclusions which can be drawn from 5.13: In any spatial region in which $\vec{v}_s$ is defined (which equals nonzero $\chi_0(\vec{r}, t)$), one gets

$$\nabla \times \vec{v}_s(\vec{r}, t) = 0, \qquad (5.14)$$

but considering a finite Path C along which $\vec{v}_s(\vec{r}, t)$ is defined enclosing a region, in which it is not, one gets the Feynman-Onsager quantization condition [67, 68]:

$$\oint \vec{v}_s dl = \frac{nh}{m} \qquad (5.15)$$

because the phase is only defined in multiples of 2π. These quantized vortices are another signature of BEC.

A different approach closely related to the definitions given before uses off-diagonal long-range order as an indicator for BEC [69]. This approach again focuses on the single-particle density matrix $\rho_1(\vec{r}, \vec{r}', t)$, but instead of explicit eigenvalues, its behavior in the limit $|\vec{r} - \vec{r}'| \to \infty$ is the subject of interest. In that case ρ_1 can instead be written in the following form:

$$\lim_{|\vec{r}-\vec{r}'|\to\infty} \rho_1(\vec{r},\vec{r}',t) = f^*(\vec{r},t)f(\vec{r}',t) + \tilde{\rho}_1(\vec{r},\vec{r}',t), \tag{5.16}$$

where $\tilde{\rho}_1(\vec{r},\vec{r}',t)$ will tend to zero for $|\vec{r} - \vec{r}'| \to \infty$. This is the behavior expected for the noncondensed fraction. $f(\vec{r},t)$ can be zero or nonzero. If it is nonzero, it can be identified with the order parameter $\psi(\vec{r},t)$ and the system is Bose-condensed. Otherwise it is not. A nonzero value of $f(\vec{r},t)$ is often identified with the presence of ODLRO. In simple cases this definition is equivalent to the definition of the order parameter in equation 5.11. It should, however, be noted that the definition using ODLRO cannot be simply applied to fragmented BECs or trapped systems, where the limit $|\vec{r} - \vec{r}'| \to \infty$ cannot be taken.

One more commonly used criterion for BEC the order parameter in terms of the Bose field operator in second quantization $\hat{\psi}(\vec{r},t)$:

$$\psi(\vec{r},t) = \langle \hat{\psi}(\vec{r},t) \rangle. \tag{5.17}$$

Here the occurrence of BEC corresponds to a non-zero value of the right-hand side of equation 5.17 analogous to the definition of a classical electric field ε_{cl} in terms of the electric field operator $\hat{\varepsilon}(\vec{r},t)$ by the prescription:

$$\varepsilon_{cl}(\vec{r},t) = \langle \hat{\varepsilon}(\vec{r},t) \rangle. \tag{5.18}$$

The field operators decrease the particle number N by one. While this is not a problem for the electromagnetic field because there are no restrictions on the field being in a superposition of states corresponding to different photon numbers, the situation is not that trivial for atoms or other massive particles. Here the right-hand side of equation 5.17 is identical to zero for any physically allowed state. One possibility to overcome this problem lies in considering spontaneously broken U(1) gauge symmetry [70]. This idea is applied in analogy to the ideal Heisenberg ferromagnet which has a Hamiltonian with O(3) rotation symmetry. Below the Curie temperature a large fractions of the spins will be aligned in parallel, but the exact direction is not specified. Under these circumstances already a small field, which vanishes in the thermodynamic limit, is enough to orient the other spins. The same explanation holds for

BEC. A small perturbation of the form

$$\hat{H}' = -(\lambda \int \hat{\psi}^\dagger(\vec{r})d\vec{r} + h.c.) \tag{5.19}$$

will cause spontaneous breaking of O(1) symmetry and result in a finite value of the order parameter. However, the validity of this approach is nontrivial. While there are plenty of small external perturbations representing external fields which can orient a ferromagnet, there is no known physical mechanism corresponding to the small perturbation given by 5.19.
Several of these criteria have already been shown to be met for polariton condensates. There have been demonstrations of macroscopic occupation of the ground state [71], build-up of spatial coherence and ODLRO across the condensate [72], observation of a Bogoliubov excitation spectrum [73], quantized vortices [74] and half vortices [75], superfluid behavior[76, 77], build-up of spontaneous polarization [78] and second-order correlation measurements [79, 80, 81]. Adding to the complex problem of identifying BEC, the existence of a critical particle density is not automatically assured. Considering the case of a parabolic dispersion, which is relevant for polaritons, one finds that n_c converges for $d > 2$, but diverges for $d \leq 2$. In two or less dimensions ODLRO cannot occur and spontaneous symmetry breaking does not exist [82, 83, 84]. However, these no-go theorems are valid only for two-dimensional systems of infinite size. For trapped systems a second-order phase transition of Kosterlitz-Thouless type is still allowed for weakly interacting bosons [85] and a superfluid state can form without the existence of strict BEC. This limitation poses fundamental conceptual questions about the validity of the BEC criteria mentioned in the overview above as many of them cannot be directly applied to trapped systems. Macroscopic population of the ground state is still a valid, but rather qualitative criterion. The first and second order coherence properties are among the best remaining signatures of BEC, but need to be handled with care: Taking the dissipative nature of cavity polaritons into account, it is clear that polaritons can never reach true equilibrium due to cavity losses, but only a nonequilibrium steady state determined by driving and decay. This places polariton BEC conceptually very close to a common photon laser which shows similar coherence properties compared to those expected for polariton BEC. Thus, this chapter will focus on identifying similarities and dissimilarities between photon lasers and polariton BEC in terms of their coherence properties.

5.3 Correlation measurements on polariton BECs and hip states

As discussed before, a signature for BEC of atoms and polaritons alike is the presence of a large number of particles sharing the same quantum state. Therefore, one might expect that the state of a polariton BEC shows the temporal coherence properties of a coherent ($g^{(2)}(0) = 1$) or N-particle number state ($g^{(2)}(0) = 1 - \frac{1}{N}$) which become indistinguishable in the limit of large occupation numbers. However, deviations from this expected value of unity have been observed for polariton condensates [79, 80, 81, 86]. The origin of this effect is not immediately clear. One of the main differences between polaritonic condensates and atomic BEC lies in the non-equilibrium nature of the polariton BEC. Attributing the deviations from full coherence to non-equilibrium features is a reasonable first assumption. Recent theoretical models [87, 88, 89] are able to reproduce the bunching effect seen for polariton BECs pumped above threshold. These models include polariton-polariton and polariton-phonon scattering as possible mechanisms to couple the condensed ground state to the uncondensed polariton population. Polariton-polariton scattering corresponds to quantum depletion of the ground state where parametric scattering of two polaritons in the ground state into two excited states with opposite momentum occurs. Due to interactions in the polariton gas, this process would take place even at absolute zero and is therefore clearly distinguishable from thermal depletion of the ground state. Thermal depletion corresponds to polariton-phonon scattering of one polariton in the ground state towards an excited state at finite momentum. Above threshold the depletion of the ground state induced by these two competing effects introduces occupation number fluctuations in the condensed fraction which result in a loss of coherence. However, as these processes depend strongly on the relative populations of the different polariton states, the polariton lifetime and the shape of the dispersion in general, a strong dependence on the detuning can be expected. Therefore, this chapter will focus on the coherence properties of the polariton condensate for various detunings. Special emphasis is put on negative detunings. With increasing negative detuning polaritons become more photonlike and acquire a shorter lifetime. At some point, usually near zero detuning, this lifetime will become shorter than the time needed for thermalization and the polariton gas will not reach self-equilibrium [90]. In this case any transition towards a degenerate state cannot be considered a thermodynamical equilibrium phase transition anymore. It is worthwhile to study whether such a state shows properties differing from that of a self-equilibrium polariton-BEC. To clearly distinguish this non-equilibrium state from common self-equilibrium polariton BECs, the former states will be referred to as highly photonic (hip) states in the following.

The sample used to study the coherence properties of polariton BECs consists comprehensively of 12 GaAs/AlAs quantum wells embedded in a planar microcavity with 16 (20) AlGaAs/AlAs mirror pairs in the top (bottom) distributed Bragg reflector. Reflectivity measurements gave results of 1.6158 eV for the bare exciton energy and 13.8 meV for the Rabi splitting. Using a Ti-Sapphire laser with a pulse duration of 1.5 ps and a repetition rate of 75.39 MHz the pump was focused to a spot approximately 30 μm in diameter on the sample at an angle of 45°. from normal incidence. The pump was resonant with the LP branch at an in-plane wavenumber of $k_\parallel = 5.8 \mu m^{-1}$ for resonant polariton injection or tuned to the first minimum of the cavity reflectivity curve at a wavelength of 744 nm for nonresonant and incoherent pumping. The emitted signal was collected using a microscope objective with a numerical aperture of 0.26. The signal was then focused on a streak camera for correlation measurements, or the Fourier plane was imaged on a monochromator for dispersion measurements.

As can be seen in figure 5.1, $g^{(2)}(0)$ does not reach the expected thermal value of 2 for any detuning as in this case there is no single-polarization fundamental mode singled out like in the measurements described before, but both realizations of the spin-degenerate ground state are detected simultaneously. Polaritons with different spin states will not interfere with each other and therefore the values of $g^{(2)}(0)$ and $g^{(3)}(0)$ expected in the thermal regime of two superposed modes are 1.5 and 3, respectively. In this figure the results for linearly polarized and circularly polarized excitation are shown in the upper and lower panel, respectively. It is striking that, if plotted on a comparable scale, $g^{(2)}(0)$ and $g^{(3)}(0)$ give results, which are in good quantitative agreement. Under linearly polarized excitation all detunings between $\Delta = +2$ meV and $\Delta = -10$ meV show a degeneracy threshold, which is evidenced by a decrease in $g^{(2)}(0)$ and $g^{(3)}(0)$ and agrees well with the position of the threshold (shown as green dashed lines in figure 5.1) evidenced in measurements of the input-output curve. Here, the threshold is defined as the point where the LP emission and the emission from the $k_\parallel = 0$ condensed state are equally strong. In the case of strong negative detuning of $\Delta = -10$ meV no differences from a common photon lasing transition are observed under linearly polarized pumping. Additional measurements of the polariton dispersion evidence that above threshold the photons are indeed emitted from the bare cavity mode in this case. Under circularly polarized pumping the threshold is not even reached for a detuning of -10 meV. This result is in accordance with previous results showing that polariton relaxation is less efficient under circularly polarized pumping, which in turn leads to a higher threshold excitation power P_{thr} [91]. For all other detunings, significant deviations from a simple photon laser behavior emerge. Even at high excitation powers the ground state

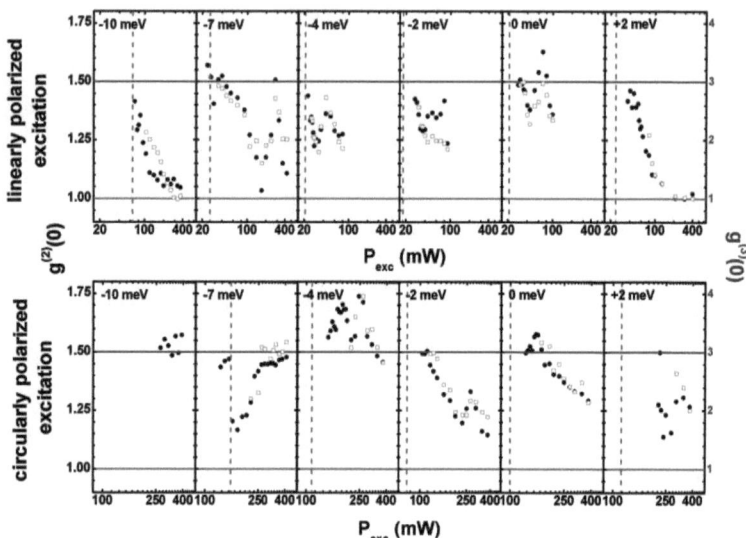

Figure 5.1: Measured $g^{(2)}(0)$ (dots) and $g^{(3)}(0)$ (open squares) determined by simultaneous two-photon and three-photon detections of the whole fundamental mode emission for a wide range of excitation powers and detunings of -10 meV ($|C|^2 \approx 77\%$), -7 meV ($|C|^2 \approx 70\%$), -4 meV ($|C|^2 \approx 62\%$), -2 meV ($|C|^2 \approx 56\%$), 0 meV ($|C|^2 \approx 50\%$) and +2 meV ($|C|^2 \approx 44\%$) under linearly polarized (upper panel) and circularly polarized (lower panel) excitation. Lower (upper) lines denote the coherent (thermal) limit. Dashed lines give the position of the degeneracy threshold determined by measurements of the dispersion.

emission has lower energy compared to the bare cavity mode with an energy difference of at least 4 meV, indicating that the strong-coupling regime is still intact. The second and third order correlation functions give further evidence that the system is not a simple photon laser in the range of detunings between +2 meV and −7 meV, corresponding to photonic contents in the range from $|C|^2 \approx 44-70\,\%$. Although at first a decay towards 1 is seen for linearly polarized excitation, especially for a detuning of −7 meV, an increase is evidenced for further increased excitation densities. Depending on the detuning, $g^{(2)}(0)$ can reach values even higher than the thermal value of 1.5. For further increased excitation power, a smooth decrease back towards 1 is observed. For circularly polarized excitation the general behavior of the correlation functions is similar to the linearly polarized case as a decrease and a reoccurence of the degenerate mode quantum fluctuations can be identified for detunings between +2 meV and −7 meV. However, in this case the correlation functions can also show increased fluctuations slightly above threshold as can be nicely seen for a detuning of −7 meV. This increase is caused by the build-up of polarization. The thermal regime value of 1.5 is just valid for unpolarized two-mode emission. As the excitation power reaches the threshold, the emission will also start to polarize and the two modes will not contribute equally to the correlation functions anymore. In the case of a superposition of two noninterfering modes A and B, the resulting measured $g^{(2)}_{AB}(0)$ is given by:

$$g^{(2)}_{AB}(0) = g^{(2)}_A(0)R_A^2 + g^{(2)}_B(0)R_B^2 + 1R_AR_B, \tag{5.20}$$

where R_A and R_B are the relative intensity ratios of mode A and B to the total intensity. Therefore it is possible to calculate $g^{(2)}_A(0)$ from $g^{(2)}_{AB}(0)$ if the relative intensity ratios and $g^{(2)}_B(0)$ are known:

$$g^{(2)}_A(0) = \frac{g^{(2)}_{AB}(0) - g^{(2)}_B(0)R_B^2 - 1R_AR_B}{R_A^2}. \tag{5.21}$$

Comparing single-mode $g^{(2)}$-measurements to the values obtained by two mode measurements shows that for circularly polarized excitation the two modes are indeed independent and the crosscircularly polarized mode stays thermal. When crossing the threshold this effect will lead to an increase in $g^{(2)}(0)$ while the buildup of coherence will lead to a decrease. Although these results are in fair qualitative agreement with mean-field and reservoir calculations of the second-order correlation function of a polariton BEC [88, 89], they are not sufficient evidence for deviations from a photon laser as the non-monotonous behavior of the correlation function can also be a result of the interplay of coherence and polarization. To make sure this is not the case, $g^{(2)}(0)$ was also studied under circularly polarized excitation for the cocircularly polarized emission component only. As shown in figure 5.2, here the expected value of $g^{(2)}(0) = 2$ is approximately reached in the limit of low excitation power for all detunings except +2 meV. At

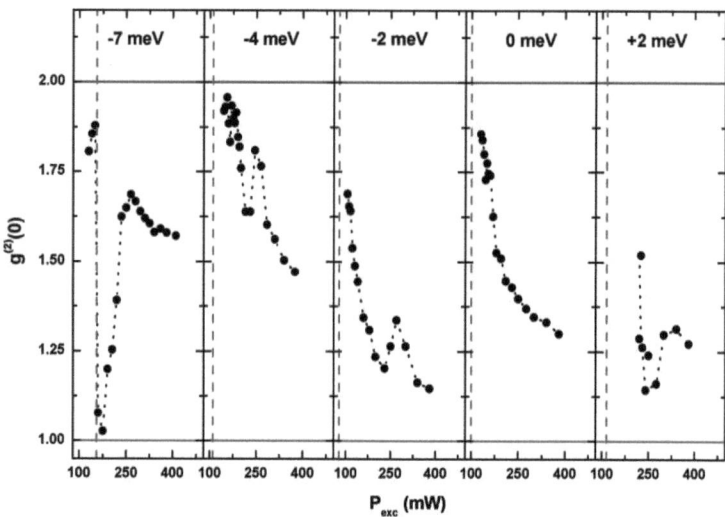

Figure 5.2: Measured $g^{(2)}(0)$ of the cocircularly polarized fundamental mode for a wide range of detunings and excitation powers under circularly polarized excitation. Lower (upper) lines denote the coherent (thermal) limit. Dashed lines give the position of the degeneracy threshold determined by measurements of the dispersion.

this detuning the emitted intensity below threshold is too small to perform sensible measurements using our setup. Above threshold the shape of $g^{(2)}(0)$ shows qualitative agreement with theoretical results [89]. For 0 meV detuning it is apparent that full coherence is not reached within the available excitation power range. Instead $g^{(2)}(0)$ decreases monotonically towards a value between 1.3 and 1.4. There is a trend towards further decrease at high excitation powers, however, the dependence on pump power is very small. Going to more negative detunings, the dip already seen without polarization sensitive detection occurs still. Apparently, the excitation power corresponding to the occurrence of the dip takes on smaller values compared to P_{thr} for larger negative detuning. However, also the rise in $g^{(2)}(0)$ seen for further increase of the excitation power increases in magnitude. This can be seen well for the most negative detuning of -7 meV where almost complete coherence is reached at an excitation power of approximately $1.1\, P_{thr}$ and a steep rise to $g^{(2)}(0) > 1.6$ is evidenced at $1.5\, P_{thr}$.

Calculations of the second-order correlation function are generally done using one of two different approaches [89]: Mean field calculations predict a decrease of $g^{(2)}(0)$ towards approximately 1.2 at the threshold, followed by a short rise for increasing excitation power until a constant value of about 1.3 is reached. This prediction is in good agreement with our results for no or small negative detuning. A two-reservoir model predicts a sharp decrease of $g^{(2)}(0)$ at the threshold, followed by a strong recurrence of the photon bunching up to values of $g^{(2)}(0) = 1.6$ and a slow drop for even higher excitation powers. This model better reproduces our results for large negative detuning. We conclude that due to the increasing relaxation bottleneck and decreased scattering rate expected for large negative detuning the two-reservoir model appears to be a valid description in the hip regime and the dip seen for several detunings is a sign of inefficient scattering between degenerate ground state polaritons and those in excited states. As the emission photon statistics depend strongly on the detuning, one might also find variations depending on the Rabi splitting and the excitonic and photonic decay constants. Therefore, microcavities operated in the hip regime may open up the possibility to introduce a high intensity light source with tunable photon statistics.

5.4 Dispersion measurements on polariton BECs and hip states

Additional characteristics of condensation and the excitation spectrum are manifested in the polariton dispersion. Above threshold a blueshift of the $k_\parallel = 0$ LP is expected as well as the LP dispersion itself is supposed to change from a parabolic shape in the uncondensed case towards a phonon-like linear dispersion in the low-momentum condensed regime $|k\xi| < 1$, where $\xi = \frac{\hbar}{\sqrt{2m_{LP}gn_c}}$ is the healing length of the condensate [73]. In standard homogeneous equilibrium Bogoliubov theory the expected dispersion is given by:

$$\omega_{Bog} = \omega_{LP} + gn_c + g_r n_r + gn_c\sqrt{(k\xi)^2((k\xi)^2 + 2)} \quad (5.22)$$

Here g and g_r are coupling constants describing the interaction between two condensate polaritons and between condensate polaritons and reservoir excitons, respectively and n_c and n_r are their densities. Using the same excitation scheme as in the previous section, the reservoir exciton contributions are expected to be negligible. Highly photonic condensed states are neither in thermal equilibrium, nor spatially homogeneous. Their spatial extent is given by the finite size of the pump spot. Therefore one might expect their dispersion to show strong deviations from the ideal homogeneous equilibrium Bogoliubov dispersion. In figure 5.3 the LP dispersion

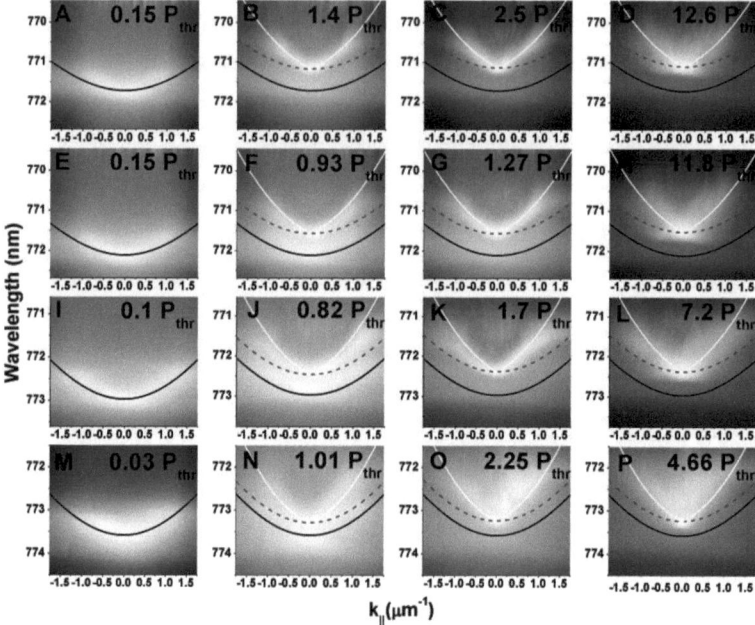

Figure 5.3: Polariton dispersions on a logarithmic scale at several excitation densities along the threshold for detunings of $\Delta = 0\,\mathrm{meV}$ (upper row), $-2\,\mathrm{meV}$ (second row), $-4\,\mathrm{meV}$ (third row) and $-7\,\mathrm{meV}$ (bottom row). Black and white lines represent the calculated LP and homogeneous equilibrium Bogoliubov dispersions, respectively. The dashed line represents the blue-shifted LP dispersion.

of the emission copolarized with the excitation is compared to the theoretical prediction for detunings of 0, -2, -4 and $-7\,\mathrm{meV}$ below and above threshold. Below threshold (A, E, I, M) the standard quadratic LP dispersion is observed in all cases. Increasing the excitation power near the threshold value (B, F, J, N), a blueshift of the $k_\| = 0$ emission energy becomes apparent, while the cross-circularly polarized emission component never shows threshold-like behavior. In addition to the blueshift of the copolarized emission also the dispersion shape starts to differ from the standard quadratic one. Blue and white lines in figure 5.3 give the quadratic LP and calculated equilibrium Bogoliubov dispersions for the condensate blueshift. For calculating ω_{Bog} the experimentally determined values of the blueshift were used for the interaction energies. It is apparent that at and above threshold the equilibrium Bogoliubov dispersion shows reasonable agreement with the measured dispersion shown for 0 and $-7\,\text{meV}$ detuning, while none of the two theoretical dispersions can reproduce the measured dispersion accurately for $k_\| = -2$ or $-4\,\text{meV}$. Here the excitation powers are slightly below threshold and the measured dispersions lie approximately in the middle between both theoretical dispersions and are not even symmetric with respect to $k_\| = 0$. The latter feature becomes more apparent with increasing negative detuning and is a signature of the nonequilibrium state favoring the presence of polaritons having a wavevector with the same sign as the pump incidence wavevector, while the first can be attributed to the inhomogeneity of the system: While the LPs with $k_\| = 0$ are stationary, LPs with $k_\| \neq 0$ will move across the excitation spot and experience a different interaction energy given by the spatial pump pulse profile. However, it is striking that the theoretical prediction again matches the experimental data well for the $\Delta = -7\,\text{meV}$ dispersion recorded at threshold. This choice of excitation power corresponds to the large dip seen in the $g^{(2)}$-measurements at the same detuning. Therefore, further theoretical research is needed to identify the microscopic mechanism connecting the suppression of occupation number fluctuations and the linearized excitation spectrum at the threshold. Further above threshold (C, G, K, O) the momentum-space region of highest intensity moves significantly closer towards $k_\| = 0$. At negative detunings there still is no complete reflection symmetry between $k_\|$ and $-k_\|$, but the positive wavevector half of the dispersion now shows reasonable agreement with the equilibrium Bogoliubov dispersion up to $k_\| \approx 0.75\,\mu\text{m}^{-1}$ without usage of any fitting parameters. This propagating sound-like mode indeed is a sign of collective behavior as expected in a condensed and thermalized polariton gas. It should be noted that the deviation of the negative wavevector half from the equilibrium Bogoliubov dispersion gets gradually larger with increased photonic content of the LP. This behavior agrees nicely with the increasing deviation from full coherence seen for hip states with large photonic content. Further increase of the excitation power (D, H, L, P) results in a more pronounced occupation of the ground state. Although the reflection symmetry between positive and negative $k_\|$ is still not perfect, both parts of the dispersion can now be described by the same equilibrium Bogoliubov dispersion with good accuracy. We interpret this behavior as a sign of effective redistribution by polariton scattering processes. Accordingly we are approaching a more thermalized, but still nonequilibrium regime for negative detunings.

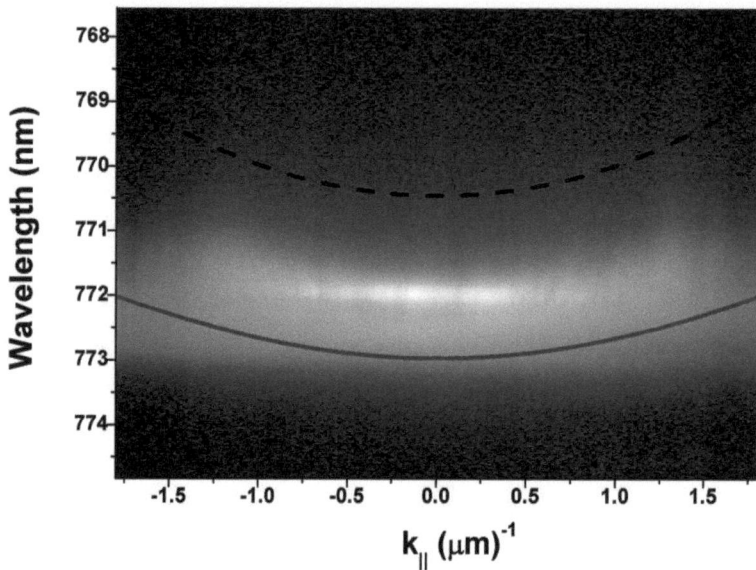

Figure 5.4: Flat polariton dispersion under nonresonant pumping at an excitation power of $P_{exc} = 12\,\text{mW}$ and a detuning of $\Delta = -4\,\text{meV}$. Solid and dashed lines denote the LP and bare cavity mode, respectively.

5.5 Nonresonantly pumped polariton BECs

Nonresonantly pumped polariton BECs show even richer physics due to their strong dependence on the pumping geometry. The exciton reservoir contributions in equation 5.22 cannot be neglected anymore, allowing for a modification of the condensate properties by changing the pumping spot size and shape [92]. The most striking consequence of this background of carriers interacting with the polariton BEC is a change of the dispersion shape. The dispersion becomes flat at some fixed energy over a wide range of $k_{||}$. Although a flat region in momentum space is predicted independently of the excitation scheme due to the diffusive nature of the Bogoliubov-Goldstone mode [93, 94], the large extent of the flat dispersion in momentum-space and the independence of this extent on the healing length of the condensate rules an interpretation in these terms out. A typical flat dispersion of a nonresonantly pumped polariton BEC at a detuning of $\Delta = -4\,\text{meV}$ is shown in figure 5.4. There are several different approaches to explain this peculiar shape of the polariton dispersion under nonresonant pumping. Some of them will be compared in this section to analyze the origin of the polariton dispersion under nonresonant pumping in III-V material systems. First, a mean-field approach given in [92] for a II-VI material system will be explained in detail. In this framework the macroscopic condensate wave function $\psi(\vec{r})$ can be obtained from a generalized nonequilibrium Gross-Pitaevskii equation (GPE) which has the following form:

$$i\hbar \frac{\partial \psi(\vec{r})}{\partial t} = \{E_0 - \frac{\hbar^2}{2m}\nabla_{\vec{r}}^2 + \frac{i\hbar}{2}[R[n_R(\vec{r})] - \gamma_c] + V_{ext}(\vec{r}) + \hbar g\left|\psi(\vec{r})\right|^2 + V_R(\vec{r})\}\psi(\vec{r}). \quad (5.23)$$

Here m denotes the effective mass of the LP, E_0 is the LP ground state energy and g gives the repulsive Coulomb interaction strength between two condensate polaritons mediated by their excitonic content [95, 96]. V_{ext} describes an external potential term representing excitonic or photonic disorder. γ_c is the polaritonic cavity loss rate. The condensate gain rate $R[n_R]$ depends on the number of noncondensed reservoir polaritons n_R [97]. The reservoir also causes a repulsive mean-field potential $V_R(\vec{r})$. In first approximation it shows a linear dependence on the reservoir density and the pump rate $P(\vec{r})$:

$$V_R(\vec{r}) \approx \hbar g_R n_R(\vec{r}) + \hbar g_P P(\vec{r}). \quad (5.24)$$

g_R and g_P are the strengths of the repulsive interactions between condensed polaritons and reservoir polaritons and free carriers, respectively. These quantities are not well known and need to be extracted from the experiment. The GPE equation 5.23 can then be combined with a rate equation for the reservoir population $n_R(\vec{r})$ as follows:

$$\dot{n}_R(\vec{r}) = P(\vec{r}) - \gamma_R n_R(\vec{r}) - R[n_R(\vec{r})]\left|\psi(\vec{r})\right|^2. \quad (5.25)$$

Reservoir polaritons decay at an effective rate γ_R which is usually large compared to the cavity decay rate. This magnitude seems a bit surprising as the pure reservoir polariton loss rate to the cavity should not be larger than the ground state loss rate, but the rapid thermalization of reservoir polaritons with a bath of high-energy carriers created by the pump beam causes the effective decay rate to be much larger. The more interesting decay channel in terms of bosonic final state stimulation is accounted for by the term proportional to $R[n_R(\vec{r})]\left|\psi\right|^2$.

The stationary solutions of equations 5.23 and 5.25 depend on the spatial shape of the pump $P(\vec{r})$. In the easiest case of a spatially homogeneous pump beam, below threshold the reservoir population grows linearly with the pump intensity as $n_R = \frac{P}{\gamma_R}$ and the condensate density is equal to zero. At the condensation threshold P_{thr} the cavity loss rate and the stimulated emission term cancel out $R[n_{R,thr}] = \gamma_c$ and the solution with zero condensate density becomes unstable. Above threshold, one gets a homogeneous, fixed reservoir density $n_R(\vec{r}) = n_{R,thr}$ and a condensate wave function of the form $\psi(\vec{r}) = \psi_0 e^{i(\vec{k}_c \vec{r} - \omega_c t)}$ leading to a condensate density of $|\psi_0|^2 = \frac{P - P_{thr}}{\gamma_c}$. However, there are several stable solutions as the value of $\vec{k}_c$ is still undetermined. For each value of $\vec{k}_c$, the corresponding condensate oscillation frequency ω_c is given by:

$$\omega_c - \omega_0 = \frac{\hbar^2 k_c^2}{2m} + g|\psi_0|^2 + g_R n_R + g_P P. \tag{5.26}$$

In the more realistic and relevant case of an inhomogeneous pump intensity profile $P(\vec{r})$, the boundary conditions at the edge of the pump spot actually fix the specific solution taken at a certain position. However, assuming a global density and phase of the condensate is not a valid approach anymore. Under these circumstances the stationary solutions will be of the form

$$\psi(\vec{r}, t) = \psi_0(\vec{r}) e^{-i\omega_c t} = \sqrt{\rho(\vec{r})} e^{i[\phi(\vec{r}) - \omega_c t]} \tag{5.27}$$

$$n_R(\vec{r}, t) = n_R(\vec{r}). \tag{5.28}$$

Here $\rho(\vec{r})$ and $\phi(\vec{r})$ represent the local density and phase of the condensate wave function, respectively. The condensate frequency ω_c is the same throughout the pumping spot, while the local condensate wave vector $\vec{k}_c(\vec{r})$ is given by the spatial gradient of the condensate phase $\vec{k}_c(\vec{r}) = \nabla_{\vec{r}} \phi(\vec{r})$ in analogy to the definition of the superfluid velocity in equation 5.13. Inserting equations 5.27 and 5.28 into equations 5.25 and 5.23, one gets the following requirements for stationary solutions:

$$\hbar \omega_c = \hbar \omega_0 + \frac{\hbar^2 k_c^2}{2m} + V_{ext} + \frac{\hbar^2}{2m} \frac{\nabla_{\vec{r}}^2 \sqrt{\rho}}{\sqrt{\rho}} + \hbar g \rho + \hbar g_R n_R + \hbar g_P P \tag{5.29}$$

$$0 = (R[n_R] - \gamma_c)\rho - \frac{\hbar}{m} \nabla_{\vec{r}} \rho \vec{k}_c \tag{5.30}$$

$$P = \gamma_R n_R + R[n_R]\rho. \tag{5.31}$$

For slowly varying pump profiles corresponding to large pump spots a local density approximation (LDA) can be applied. Under these circumstances the quantum pressure term in equation 5.29 and the current divergence term in 5.30 can be neglected. In this approach, the condensate is treated as a homogeneous system with a local value of the pump intensity. This also means that $\rho(\vec{r}) = 0$ for all points $\vec{r}$ where the pump intensity is below the threshold. Experiments show that in the area where a condensate is still present, ω_c stays constant in space. Therefore the variation of $P(\vec{r})$ along the pump spot needs to be compensated by a spatial variation of $\vec{k}_c(\vec{r})$. In the simple case of a circular spot $P(\vec{r}) = P(r)$ analytical solutions can be found if no disorder is present. In this case the stationary solutions need to be cylindrically symmetric and the local wave vector at the center of the spot vanishes: $\vec{k}_c(r=0) = \vec{0}$. This results in:

$$\omega_c - \omega_0 = g\rho(r=0) + g_R n_R(r=0) + g_P P(r=0). \tag{5.32}$$

For common Gaussian pump spots the local wave vector $\vec{k}_c$ is always pointing in the radial direction and its modulus increases with r. Its largest value is reached at the condensate border where the pump intensity equals P_{thr}. In this case the repulsive interactions can be considered an antitrapping potential

$$V_{at} = \hbar g\rho(r) + \hbar g_R n_R(r) + \hbar g_P P(r), \tag{5.33}$$

which causes the polaritons to be accelerated away from the center.
Calculated results for the polariton distribution in real space, the emitted intensity in momentum space and the local wavevector $\vec{k}_c(r)$ in the case of a constant pump rate and a circular spot with a diameter of $18\,\mu m$ are shown in figure 5.5. The real space polariton distribution in the upper panel shows a clear broadening due to the antitrapping potential. The local polariton number vanishes at the position where the polariton density falls below the threshold polariton density. At this point also the local condensate wave vector shown in the middle panel starts to saturate. The rather steep slope of $\vec{k}_c(r)$ near the center of the pump spot results in a broad polariton distribution in momentum space. It is contained within the region where the condensate energy is larger than the quadratic free particle dispersion as seen in the bottom panel. These results can explain the flat dispersions seen in II-VI polariton OPOs [98] and nonresonantly pumped II-VI polariton condensates using large pumping spots [99] well. However, the momentum space distribution of III-V based polariton condensates behaves differently and calls for a different model. An example is shown in the upper panel of figure 5.7 where the intensity distribution in momentum space for a III-V based polariton condensate at zero detuning with $18\,\mu m$ pumping spot size is compared to the theoretical calculations based on a parabolic polariton dispersion. Obviously the broad distribution predicted for this pump geometry does not occur. Instead a narrower and peaked distribution with a superimposed substructure occurs. This substructure is a consequence of a slight disorder potential present in the system. However, the dicrepancy between the experimental data and the theoretical prediction calls for a modified theoretical model. The two main differences between the theoretical assumptions and the experimental conditions are the usage of a III-V system and pulsed excitation. Also, the above model does not consider the thermodynamic properties of the polariton BEC. Therefore, a first possible modification lies in the particular shape of the underlying polariton dispersion. The above approach assumes a parabolic LP-like dispersion

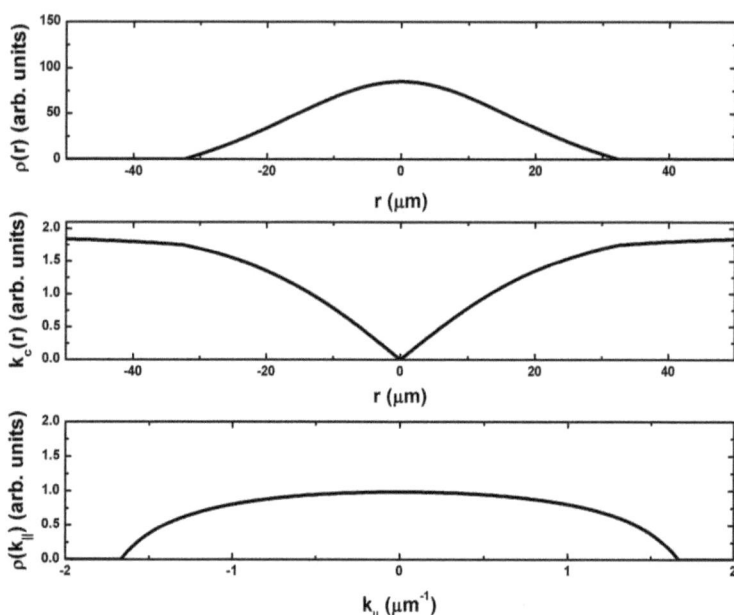

Figure 5.5: Calculated results using a LDA approach for a pumping spot of 18 µm diameter assuming an underlying parabolic dispersion. The upper panel gives the spatial polariton density, the middle panel gives the local condensate wave vector $\vec{k}_c(r)$ and the bottom panel gives the emitted intensity in momentum space. The intensity is calculated from the polariton distribution in momentum space considering the varying lifetimes of polaritons with different $k_\parallel$.

for the polaritons. In III-V based cavities, however, the presence of a linearized Bogoliubov dispersion has been demonstrated [73] as also shown in section 5.4. Therefore, it seems reasonable to calculate the local condensate wave vector and the resulting intensity distribution assuming a modified dispersion corresponding to equation 5.22. Results calculated for the case of a linearized dispersion, but keeping all other constants the same as before are shown in figure 5.6. As seen in the upper panel, the local condensate wavevector does not increase as steep

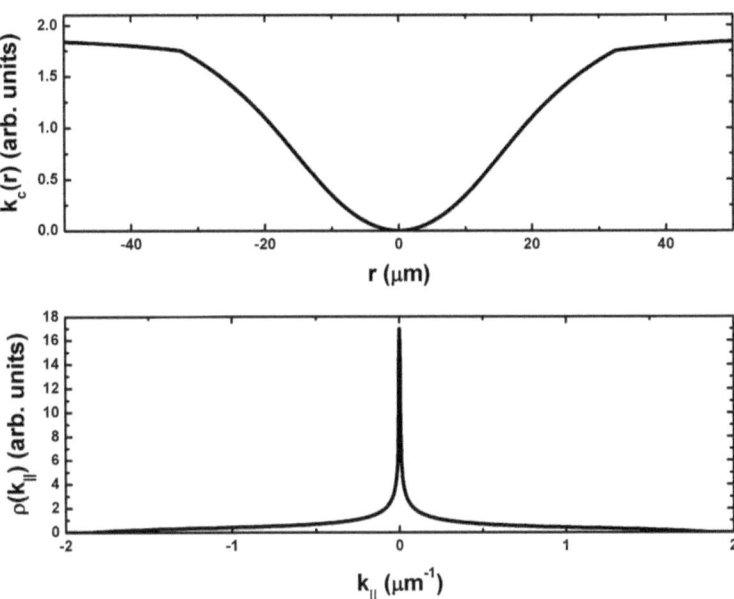

Figure 5.6: Calculated results using a LDA approach for a pumping spot of $18\,\mu$m diameter assuming an underlying Bogoliubov dispersion. The upper panel gives the spatial polariton density, the middle panel gives the local condensate wave vector $\vec{k}_c(r)$ and the bottom panel gives the emitted intensity in momentum space. The intensity is calculated from the polariton distribution in momentum space considering the varying lifetimes of polaritons with different $k_{||}$.

as before at positions away from the center of the pump beam. As the Bogoliubov dispersion is steeper than the parabolic one, a smaller $\vec{k}_c(r)$ is sufficient to level the energies at different positions throughout the pumping spot and $\vec{k}_c(r)$ shows an almost parabolic dependence on the position. At large distances from the center of the pump spot the local polariton density decreases and the dispersion shows deviations from linear behavior and $\vec{k}_c(r)$ saturates. Accordingly, the corresponding momentum-space emission pattern is much narrower compared to the one for a parabolic dispersion and most of the intensity will be emitted at small wave vectors. However, as shown in the second panel of figure 5.7 this model is also not sufficient to explain the emission pattern seen in experiments which is broader in momentum space. A broadening due to the limited resolution of the measurement apparatus alone is not sufficient to explain this difference. Another possibility to explain the emission pattern can be given by taking redistribution in terms of polariton-polariton scattering into account. This mechanism conserves the total momentum of the scattering particles. For the ground state this results in two particles with $k_\parallel = 0$ scattering into states with opposite momentum and results in depletion of the ground state [73]. Also thermally activated excitations from the ground state are possible.

In the following a toy model will be used to investigate this approach. It is based on a semi-classical Boltzmann equation treatment. The populations n_k of the states of the condensate dispersion are assumed to follow bosonic coupled rate equations of the following form:

$$\frac{dn_k}{dt} = P_k - \gamma_k n_k - n_k \sum_{k'} W_{k \to k'}(1 + n_{k'}) + (1 + n_k) \sum_{k'} W_{k' \to k} n_{k'}. \qquad (5.34)$$

While this set of coupled rate equations looks simple, calculating the scattering rates is a nontrivial task. The easiest approach lies in using the Fermi golden rule. However, such a method requires the possibility of treating the scattering processes in a perturbative manner. In the strong coupling regime and especially in the Bogoliubov regime massive energy renormalization is present and a completely perturbative treatment is not possible. Therefore, a non-perturbative treatment of these phenomena is necessary. Here this problem is solved by treating all non-perturbative effects in terms of a modified dispersion. The initial distribution of the polaritons in momentum space is then assumed to follow the Bogoliubov dispersion as shown in the lower panel of figure 5.6 and their interactions will only cause scattering within this dispersion which can be calculated perturbatively via the Fermi golden rule. The most important contribution to the polariton distribution will be caused by polariton-polariton scattering. The polariton-polariton scattering rate is given by:

$$W^{pol}_{k \to k'} = \frac{2\pi}{\hbar} \sum_q \frac{N_q(1 + N_{q+k'-k}) \left|M^{X-X}\right|^2 |X_k|^2 |X_{k'}|^2 |X_q|^2 |X_{q+k'-k}|^2 \gamma_{k'}}{(E(k') - E(k) + E(q + k' - k) - E(q))^2 + \gamma_{k'}^2}. \qquad (5.35)$$

It depends on the polariton linewidths γ_k, the excitonic Hopfield coefficients, the exciton-exciton interaction matrix element M^{X-X} and the instantaneous population of all other polariton modes. The exact value of M^{X-X} is difficult to determine [100, 101]. The following numerical estimate will be used [54] which is sufficient for the toy model presented here:

$$M^{X-X} \approx 6\frac{a_B^2}{S}E_b. \quad (5.36)$$

a_B gives the two-dimensional Bohr radius [102], E_b is the exciton binding energy and S gives the normalization area. The differential equations are evaluated on a linear grid in momentum space. Cylindrical symmetry is assumed and well justified in the case of non-resonant pumping. Therefore, the scattering rates from all states belonging to one point of the grid to all states belonging to one other point of the grid are equal. The initial occupation of the polariton states was calculated from the emitted intensity pattern assuming an occupation on the order of unity at the border of the condensate and takes the polariton lifetimes and the density of states into account. However, the initial values can at best be a rough estimate. Simulations using this model show a rapid redistribution of polaritons in momentum space away from the ground state for large initial polariton numbers. At large times the ground state is depleted and most of the emission comes from states with large $k_{||}$. The time-integrated emission of the condensate is shown in the third panel of figure 5.7. While this model gives a better estimate for the intensity distribution near the ground state, it predicts a redistribution of the polaritons which is much larger than seen in the experiment and results in an emission pattern which is too broad.

It should be noted that the aforementioned models still treat the effect of condensation as some kind of boundary condition causing polaritons to have some common energy or phase, but still treat the polaritons as some kind of individual particles. A final possible modification lies in explicitly treating the condensate as a single state with a population of the excitation spectrum depending on the thermodynamic properties of the condensate [73]. Considering the condensate as a macroscopic state. For this model the excitation spectrum is again assumed to follow a linearized dispersion as given by equation 5.22. The average population of Bogoliubov particles with a fixed $k_{||}$ is then simply given by:

$$\langle \hat{v}^\dagger_{k_{||}} \hat{v}_{k_{||}} \rangle = \frac{1}{e^{\frac{E_{Bog}}{k_B T}} - 1}. \quad (5.37)$$

However, the particles actually seen in the experiments are not the Bogoliubov particles, but the lower polaritons. Accordingly, a Bogoliubov transformation can be used to calculate the

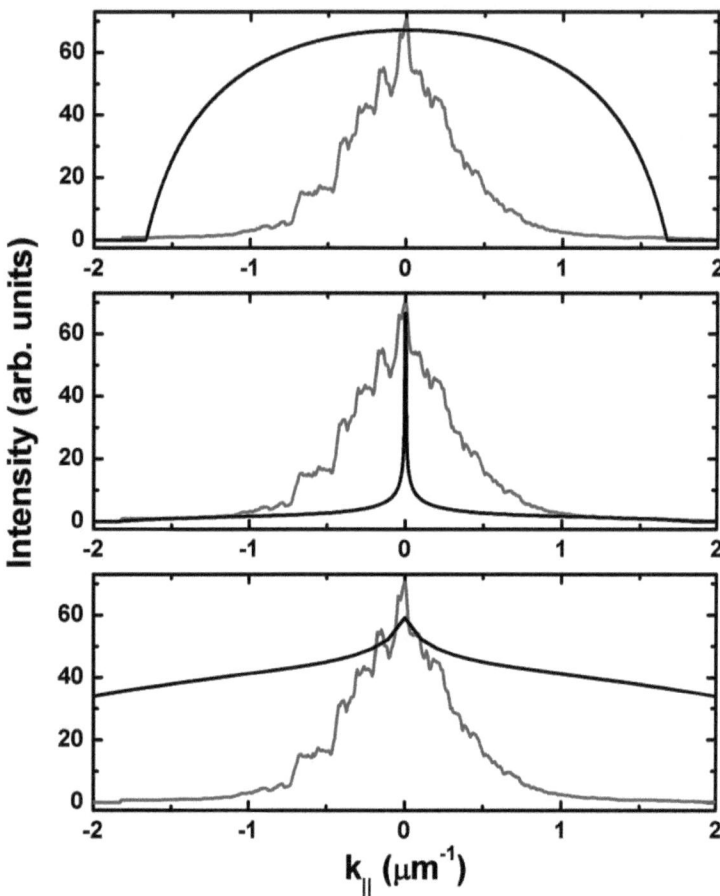

Figure 5.7: Expected momentum-space intensity distributions of a nonresonantly pumped polariton BEC and a pump diameter of $18\,\mu m$ compared for several common models. The upper panel gives the results of a mean-field approach assuming an underlying parabolic excitation spectrum. The middle panel uses the same approach, but assumes a linearized Bogoliubov dispersion. The lower panel shows the prediction using the linearized model as a starting point and considering redistribution in terms of stimulated scattering processes. All models give wrong results for large regions of the dispersion.

expected distribution in terms of polaritons and emitted intensity:

$$\hat{q}_{k_\|} = f_{k_\|}\hat{v}_{k_\|} + g^*_{-k_\|}\hat{v}^\dagger_{-k_\|} \tag{5.38}$$

$$\hat{q}^\dagger_{k_\|} = f^*_{k_\|}\hat{v}^\dagger_{k_\|} + g_{-k_\|}\hat{v}_{-k_\|} \tag{5.39}$$

$$f_{k_\|}, g_{-k_\|} = \pm\sqrt{\frac{\hbar^2 k_\|^2}{4m_{LP}E_{Bog}} + \frac{\hbar g n_c}{2E_{Bog}} \pm \frac{1}{2}}. \tag{5.40}$$

The occupation number of the observable particles is then given by:

$$n_{k_\|} = \langle \hat{q}^\dagger_{k_\|}\hat{q}_{k_\|}\rangle = \left|f_{k_\|}\right|^2 \langle \hat{v}^\dagger_{k_\|}\hat{v}_{k_\|}\rangle + \left|g_{-k_\|}\right|^2 (1 + \langle \hat{v}^\dagger_{k_\|}\hat{v}_{k_\|}\rangle) \tag{5.41}$$

$$= \left|g_{-k_\|}\right|^2 + \frac{\left|f_{k_\|}\right|^2 + \left|g_{-k_\|}\right|^2}{e^{\frac{E_{Bog}}{k_B T}} - 1}. \tag{5.42}$$

This occupation number consists of one temperature-dependent and one temperature-independent term. The temperature independent term $\left|g_{-k_\|}\right|^2$ causes particles to be emitted at finite $k_\|$ even at absolute zero, where $\langle \hat{v}^\dagger_{k_\|}\hat{v}_{k_\|}\rangle = 0$. This mechanism is called quantum depletion and corresponds to the simultaneous spontaneous excitation of two polaritons in the ground state towards states with opposite $k_\|$. The other term represents thermal depletion and corresponds to the thermal excitation of one polariton from the ground state towards some state with finite $k_\|$. Both mechanisms can be distinguished by taking their dependence on $k_\|$ into account. At small $k_\|$ one can assume terms of order $k\xi$ to be dominant, while terms of order $(k\xi)^2$ will vanish. In this limit the occupation numbers according to quantum depletion will behave as follows:

$$\langle \hat{q}^\dagger_{k_\|}\hat{q}_{k_\|}\rangle \approx \frac{1}{2\sqrt{2}k\xi}. \tag{5.43}$$

In the case of thermal depletion at small $k_\|$, $\left|f_{k_\|}\right|^2 + \left|g_{-k_\|}\right|^2$ can be approximated as $\frac{\hbar g n_c}{E_{Bog}}$ and in the low-energy regime the Bogoliubov particle occupation can be approximated as $\langle \hat{v}^\dagger_{k_\|}\hat{v}_{k_\|}\rangle \approx \frac{k_B T}{E_{Bog}}$. Neglecting terms of order $\hbar^4 k_\|^4$, the occupation numbers in the case of thermal depletion are given approximately by:

$$\langle \hat{q}^\dagger_{k_\|}\hat{q}_{k_\|}\rangle \approx \frac{m k_B T}{\hbar^2 k_\|^2}. \tag{5.44}$$

Therefore, the emitted intensity should either show a $k_\|^{-1}$ or $k_\|^{-2}$ dependence.

The experimental data is compared to these predictions in a double-logarithmic plot in figure 5.8. The differing photonic fractions at different $k_\|$ have been considered. Red and blue lines give the expected emitted intensity for thermal and quantum depletion, respectively. Even for the low temperature of 10 K used in the experiment, thermal depletion is dominant by at least a factor of four for large $k_\|$ and even by more than one order of magnitude for small $k_\|$.

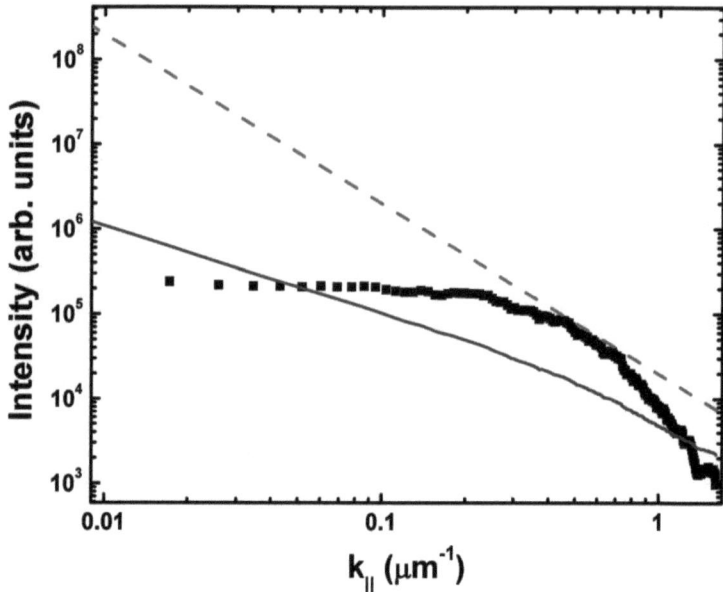

Figure 5.8: Theoretical predictions of polariton occupation numbers for quantum (solid line) and thermal (dashed line) depletion approximated for small $k_\|$. Black squares give the experimental results. Thermal depletion is always dominant, but can explain the observed distribution only in a very narrow range.

However, thermal depletion can reproduce the distribution seen in the experiment only for a narrow range of wave vectors between $0.4\,\mu\text{m}^{-1}$ and $0.75\,\mu\text{m}^{-1}$. For smaller and larger wave vectors, the theoretical prediction gives too large values by more than one order of magnitude. The discrepancy at low wave vectors can be explained by taking the finite size of the pumping spot into account. As the condensate is contained within the finite spot size R, it is in some respects comparable to a trapped condensate with the steep decrease of the excitation power in real space acting as the trapping potential. However, this potential still differs significantly from a potential well as there is a shallow decrease of the excitation power past the steep one. In theoretical treatments this trapping potential is often modeled by considering a finite spacing of $2\pi/R$ between the ground state and the first excited state, while the remaining states in reciprocal space are assumed to vary continuously [16]. For the case of a 18 μm pump spot the distance between the steepest edges of the pulse shape will be somewhere between 14 and 18 μm, corresponding to a ground state width between 0.35 μm^{-1} and 0.45 μm^{-1}. These agrees remarkably well with the region in figure 5.8 up to $k_{||} = 0.4\,\mu m^{-1}$ inside which the experimental data shows only a small decrease with $k_{||}$. As the basic Bogoliubov theory used to predict the amount of thermal depletion assumes an untrapped condensate, the results for wavevectors below the continuum of states are very inaccurate. Interpreting the emission from these small wave vectors as belonging to a single quantum state gives a better explanation in this region. The deviations at larger wave vectors require a more detailed explanation. Strictly speaking, equations 5.43 and 5.44 are valid only in the regime of small wave vectors, where the dependence of the condensate depletion on the local polariton density can be neglected. At larger wave vectors, these contributions become important and the complete expressions for quantum and thermal depletion are given by

$$\langle \hat{q}^\dagger_{k_{||}} \hat{q}_{k_{||}} \rangle \propto \frac{\frac{\hbar^2 k_{||}^2}{2m} + \hbar g n_c}{2 E_{Bog}} - \frac{1}{2} \quad (5.45)$$

and

$$\langle \hat{q}^\dagger_{k_{||}} \hat{q}_{k_{||}} \rangle \propto \frac{2 m k_B T}{\hbar^2 k_{||}^2 (\frac{\hbar^2 k_{||}^2}{2m \hbar g n_c} + 2)}, \quad (5.46)$$

respectively. Both relations require knowledge of the underlying polariton-polariton interaction energy $U = \hbar g n_c$ at each position. Evaluating these in the case of non-resonant pumping is a complicated task as the polariton-polariton interaction is not the dominant source of the condensate blueshift as in the case of resonant pumping. As a rather qualitative model, one can assume the polariton-polariton interaction energy at the degeneracy threshold to equal the condensate blueshift at the threshold seen for resonant pumping. In both cases degeneracy should set in at a ground state population of unity. In the resonant pumping scheme the polariton-polariton interaction is the only important contribution to the blueshift. Therefore, the interaction energies should be of the same order. In this case, a value of $U_0 = \hbar g n_c = 1.2\,\text{meV}$ is assumed at the center of the pump spot. The local interaction constants at other positions are then given by the relative occupations at the different positions and scale as $U_0 \frac{n(\vec{r})}{n_0}$. However, finding the equivalent values in momentum space is more demanding. The polariton occupation in k-space alone is not a sensible estimate as emission with a certain $k_{||}$ can come from different positions inside the pumping spot and the corresponding polaritons will move through several regions with different polariton densities during their life time. As polaritons with finite $k_{||}$ usually move away from the center of the pump spot, the mean polariton density in their vicinity will be higher than the one given by the occupation number at $k_\parallel$. A sensible approach lies in assuming local condensate wave vectors $k_c(r)$ like in the first approach used again and using an averaged polariton density giving the average over all positions between the center of the pumps spot and the position where $k_\parallel = k_c(r)$. The resulting function is nontrivial, but can reasonably well be approximated by

$$U(k) = U(k=0)\sqrt{\frac{n(k)}{n(k=0)}}. \tag{5.47}$$

Inserting these results in the expressions for quantum and thermal depletion yields distributions as shown by solid blue and red lines in figure 5.9. Dotted lines give the corresponding approximations neglecting higher-order terms. Thermal depletion is still dominant over quantum depletion by at least one order of magnitude. Additionally, a decrease of thermal depletion at large wave vectors is seen. This behavior reproduces the values seen in the experiment well in the region beyond the condensate ground state. This is a good argument for the elementary underlying excitation spectrum of a nonresonantly pumped polariton condensate being of a linearized Bogoliubov kind, while the additional blue shift causing the flat dispersion seen in experiment stems from a background of carriers created by the pump pulse. However, whether this underlying dispersion is strictly linear as claimed in [73] or rather of a diffusive and flat nature at small $k_\parallel$ as predicted in [93] cannot be finally answered at this point, although the flat intensity distribution within the condensate ground state seems to suggest the existence of a diffusive region. Further experimental and theoretical investigations, for example in terms of a varying pump spot diameter, are needed to clear this point up.

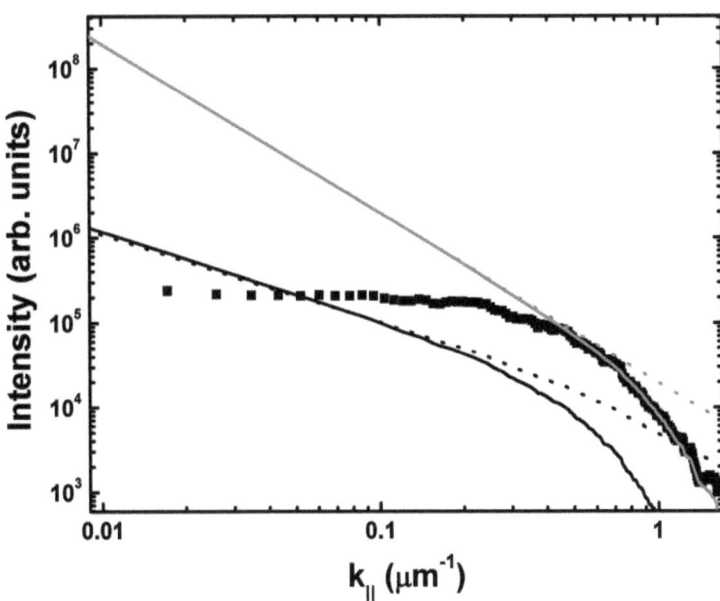

Figure 5.9: Theoretical predictions of polariton occupation numbers for quantum (dark line) and thermal (bright line) depletion when taking higher order contributions into consideration. For comparison, the predictions neglecting these contributions are shown as dotted lines. Thermal depletion is the dominant mechanism populating states at finite $k_{||}$. The experimental data is well reproduced for wave vectors beyond the extent of the condensate ground state.

Chapter 6

Summary and Outlook

This thesis introduced and characterized a novel intensity correlation technique. By using a streak camera instead of photodiodes a time resolution on the order of 2 ps was realized. This technique was used to access time-resolved correlations inside short pulses emitted from semiconductor light sources. Thereby it was possible to identify the transition from spontaneous to stimulated emission in these structures. In detail, it was possible to identify the transition from spontaneous to stimulated emission of the emission from quantum dot micropillar lasers in a time integrated as well as two different time resolved manners and reveal a region of nonclassical light emission. Further studies on QW microcavity systems allowed us to trace the bleaching of the strong coupling regime accompanied with the onset of lasing in these structures and its reflection in the second- and higher order correlation functions. A different sample made consisting of 12 QWs embedded in a microcavity made it possible to investigate a system where condensation occurs before the strong coupling regime is bleached. Here it was possible to demonstrate that this special kind of non-equilibrium condensate shows deviations from ideal coherence seen for atomic condensates. For resonant excitation under large angles, these deviations are caused by scattering to and from the condensate ground state and were shown to be strongly dependent on the detuning between the exciton and the cavity mode. For nonresonant excitation a clear description of the condensed state showed to be more complicated as the excitation spectrum of the condensate is not easily accessible because strong interactions with background carriers are dominant. Nevertheless, it could be shown that the momentum distribution of the emitted intensity gives evidence for a linearized Bogoliubov excitation spectrum with occupation numbers based mainly on the thermodynamical properties of the system.

From here on the work can be continued in several directions. From a technical point of view, there are still several possibilities to optimize the efficiency of the streak camera approach for given experimental situations. Especially, the repetition rate of the CCD used is a limiting factor. Replacing it by an array of avalanche photodiodes connected to on-board logical circuits realized by field programmable gate arrays could lead to an enormous increase in the possible readout rate and drastically shorten the necessary integration times. It might therefore become possible to analyze the photon statistics of weak emitters, down to the level of single photon sources using the presented approach. Also, at present the streak camera correlation technique can only be applied to pulsed signals at the repetition frequency of the laser triggering the streak camera, but not for CW signals. However, it is in principle possible to use a pulsed signal for triggering the camera, but a CW laser for performing the experiment. To realize this case, the photo cathode off the streak camera should not receive any signal while the horizontal deflection unit is gated off. Otherwise the large constant optical intensity might induce irreparable damage to the photocathode. This could in principle be achieved by feeding the CW signal into an acousto-optic modulator triggered by the same signal used to trigger the horizontal deflection unit. Doing so would create long pulses with constant intensity over a duration of approximately 600 ns, which is longer than the coherence times of interest by a factor of 10^4 and can be considered as pseudo-CW excitation. From a more physical point of view, there are still plenty of open questions. The microscopic origin of the non-classical light emission seen for the high-Q QD micropillar sample is still not completely clear. Further, it might be interesting to study other ultrafast phenomena like superradiant light emission of quantum dot ensembles. Also the field of polaritonic BECs still offers plenty of further research directions. Whether the photon statistics seen in the case of resonant excitation under large angles will be reproduced in the case of nonresonant pumping or pumping at the magic angle, is not clear. Also, the modified inscattering and outscattering rates of the condensate ground state due to acoustic phonon scattering at elevated temperatures might result in modified intensity correlation properties. Finally, it might be worthwhile to study the modified properties of the polariton condensate under the influence of an external magnetic field, where phenomena like the polariton counterpart of the Spin-Meissner effect have been predicted [103].

Appendix A

Theoretical Model of QD Micropillar Lasers

The microscopic QD laser theory applied throughout this thesis was developed in the semiconductor theory group of Prof. Jahnke at the University of Bremen [35, 104]. It makes use of the cluster expansion technique [105, 106]. The light field and the carrier system are treated in second quantization. The corresponding total Hamiltonian reads:

$$\hat{H} = \hat{H}_0^{carr} + \hat{H}_{Coul} + \hat{H}_{ph} + \hat{H}_D. \tag{A.1}$$

It consists of four contributions. The carrier part contains the single-particle contributions for conduction and valence band carriers with energy $\varepsilon_\nu^{c,v}$:

$$\hat{H}_0^{carr} = \sum_\nu \varepsilon_\nu^c \hat{c}_\nu^\dagger \hat{c}_\nu + \sum_\nu \varepsilon_\nu^v \hat{v}_\nu^\dagger \hat{v}_\nu. \tag{A.2}$$

$\hat{c}_\nu$ and $\hat{c}_\nu^\dagger$ are fermionic annihilation and creation operators for conduction band carriers in the state $|\nu\rangle$. $\hat{v}_\nu$ and $\hat{v}_\nu^\dagger$ are the corresponding operators for valence band carriers. The Hamiltonian describing the two-particle Coulomb interaction with Coulomb matrix elements $V_{\alpha'\nu\nu'\alpha}^{\lambda\lambda'}$ is given by:

$$\hat{H}_{Coul} = \frac{1}{2} \sum_{\alpha'\nu\nu'\alpha} \left[V_{\alpha'\nu,\nu'\alpha}^{cc} \hat{c}_{\alpha'}^\dagger \hat{c}_\nu^\dagger \hat{c}_{\nu'} \hat{c}_\alpha + V_{\alpha'\nu,\nu'\alpha}^{vv} \hat{v}_{\alpha'}^\dagger \hat{v}_\nu^\dagger \hat{v}_{\nu'} \hat{v}_\alpha \right] + \sum_{\alpha'\nu\nu'\alpha} V_{\alpha'\nu,\nu'\alpha}^{cv} \hat{c}_{\alpha'}^\dagger \hat{v}_\nu^\dagger \hat{v}_{\nu'} \hat{c}_\alpha. \tag{A.3}$$

The electromagnetic field Hamiltonian has the form

$$\hat{H}_{ph} = \sum_q \hbar\omega_q \left(\hat{b}_q^\dagger \hat{b}_q + \frac{1}{2} \right), \tag{A.4}$$

where $\hat{b}_q$ and $\hat{b}_q^\dagger$ are bosonic operators annihilating and creating a photon in mode q. Light-matter interaction is considered in dipole approximation. The corresponding two-particle

Hamiltonian is given by:

$$\hat{H}_D = -i \sum_{q,\alpha\nu} \left(g_{q\alpha\nu} \hat{c}_\alpha^\dagger \hat{v}_\nu \hat{b}_q + g_{q\alpha\nu} \hat{v}_\alpha^\dagger \hat{c}_\nu \hat{b}_q \right) + H.c., \tag{A.5}$$

where the light-matter coupling strength $g_{q\alpha\nu}$ is determined by the overlap of the electromagnetic field mode function with index q and the single-particle wave functions corresponding to states $|\alpha\rangle$ and $|\nu\rangle$.

Starting from the full Hamiltonian $\hat{H}$, coupled equations of motion for the carrier dynamics and the temporal behavior of the photon modes can be derived in the Heisenberg picture. The occurring operator averages for the different quantities of interest can now be classified according to the number of particles involved. For example, the electron and hole populations $f_\nu^e = \langle \hat{c}_n u^\dagger \hat{c}_n u \rangle$, $f_\nu^h = 1 - \langle \hat{v}_n u^\dagger \hat{v}_n u \rangle$ are singlet contributions, the source term of spontaneous emission $\langle \hat{c}_\alpha^\dagger \hat{v}_\alpha \hat{v}_\nu^\dagger \hat{c}_\nu \rangle$ and the photon-assisted polarization $\langle \hat{b}_q^\dagger \hat{v}_\nu^\dagger \hat{c}_\nu \rangle$ are doublet terms.

It is possible to describe these N-particle averages $\langle N \rangle$ in a factorized sum of one- up to (N-1)-particle averages. The difference between the factorization and the full average can be expressed as a correlation function of order N, denoted as $\delta\langle N \rangle$. Accordingly factorizations are given by

$$\langle 1 \rangle = \delta\langle 1 \rangle, \tag{A.6a}$$

$$\langle 2 \rangle = \langle 1 \rangle\langle 1 \rangle + \delta\langle 2 \rangle, \tag{A.6b}$$

$$\langle 3 \rangle = \langle 1 \rangle\langle 1 \rangle\langle 1 \rangle + \langle 1 \rangle\delta\langle 2 \rangle + \delta\langle 3 \rangle, \tag{A.6c}$$

$$\langle 4 \rangle = \langle 1 \rangle\langle 1 \rangle\langle 1 \rangle\langle 1 \rangle + \langle 1 \rangle\langle 1 \rangle\delta\langle 2 \rangle + \langle 1 \rangle\delta\langle 3 \rangle + \delta\langle 4 \rangle, \tag{A.6d}$$

and so on. The two-particle parts of the Hamiltonians $\hat{H}_{Coul}$ and $\hat{H}_D$ cause an infinte hierarchy when one tries to set up equations of motion in the Heisenberg picture. The basic idea of the cluster expansion method lies in replacing occurring operator expectation values $\langle N \rangle$ according to equations A.6, so that instead equations of motion for the correlation functions $\delta\langle N \rangle$ are obtained. It is then possible to truncate the hierarchy of correlation functions at a desired level and allows to include correlations consistently up to the chosen order in all appearing operator expectation values. Truncating the hierarchy at the doublet level leads to the so-called semiconductor luminescence equations which has been successfully applied to the luminescence dynamics of quantum wells [107] and QDs [108, 109]. Addressing photon correlations requires truncation at the quadruplet level.

The QD-laser model does not consider every cavity mode individually, but distinguishes between one dominant lasing mode of photon number $\langle \hat{b}^\dagger \hat{b} \rangle$ where the index is omitted and all non-lasing modes with index nl, allowing to define the β-factor in accordance with equation 3.1. The lasing

transition is assumed to be resonant with the radiative s-shell electron-hole recombination of a QD ensemble. The pumping process is modeled as resonant optical excitation of the p-shell at a constant pumping rate P and takes saturation effects caused by Pauli blocking into account. Carrier relaxation from the p-shell to the s-shell is considered in relaxation-time approximation [110] where different relaxation times $\tau_r^{e,h}$ are used for electrons and holes. The cavity loss rate $2\kappa_q$ can be directly determined from the cavity mode Q-factor $Q = \hbar\omega/2\kappa_q$.
In the framework of this model, the evolution of the occupancy of a cavity mode q and the carrier populations are given by:

$$(\hbar\frac{d}{dt} + 2\kappa_q)\langle \hat{b}_q^\dagger \hat{b}_q \rangle = 2\Re(\sum_\nu |g_{q\nu}|^2 \langle \hat{b}_q^\dagger \hat{v}_\nu^\dagger \hat{c}_\nu \rangle). \qquad (A.7)$$

This dependence points out the major role of the photon-assisted polarization $\langle \hat{b}_q^\dagger \hat{v}_\nu^\dagger \hat{c}_\nu \rangle$ which is basically the expectation value for creation of a photon combined with an interband transition of an electron from the conduction to the valence band.in the photon number dynamics. The sum over ν involves all possible interband transitions from several QDs. The photon-assisted transition amplitude becomes damped due to carrier-carrier and carrier-phonon interactions. This behavior is reflected by introducing a phenomenological damping constant Γ. Taking Pauli blocking and the coupling to the intracavity photon number into account, the corresponding equation of motion for the most important photon-assisted polarization $\langle \hat{b}^\dagger \hat{v}_s^\dagger \hat{c}_s \rangle$ involving the s-shell and the lasing mode is then given by:

$$(\hbar\frac{d}{dt} + \kappa + \Gamma)\langle \hat{b}^\dagger \hat{v}_s^\dagger \hat{c}_s \rangle = f_s^e f_s^h - (1 - f_s^e - f_s^h)\langle \hat{b}^\dagger \hat{b} \rangle + \delta\langle \hat{b}^\dagger b \hat{c}_s^\dagger \hat{c}_s \rangle - \delta\langle \hat{b}^\dagger b \hat{v}_s^\dagger \hat{v}_s \rangle, \qquad (A.8)$$

which in turn depends on the s-shell carrier dynamics $f_s^{e,h}$. As the relaxation processes from the p- to the s-shell are included in relaxation-time approximation at a rate $R_{p\to s}^{e,h} = (1 - f_s^{e,h})\frac{f_p^{e,h}}{\tau_r^{e,h}}$, the equations of motion for the carrier dynamics are given by

$$\frac{d}{dt}f_s^{e,h} = -2|g_{q_l,s}|^2 \Re\langle \hat{b}^\dagger \hat{v}_s^\dagger \hat{c}_s \rangle - (1-\beta)\frac{f_s^e f_s^h}{\tau_{sp}} + R_{p\to s}^{e,h} \qquad (A.9a)$$

$$\frac{d}{dt}f_p^{e,h} = P(1 - f_p^e - f_p^h) - \frac{f_p^e f_p^h}{\tau_{sp}^p} - R_{p\to s}^{e,h}, \qquad (A.9b)$$

respectively. The separate inclusion of electron and hole dynamics represents one of the major difference to the situation in atomic approaches. This set of equations already allows to determine the temporal evolution of carrier and photon number dynamics. Accessing photon number correlations requires consideration of quadruplet contributions, too. The equal-time correlation function can be calculated as

$$g^{(2)}(0) = \frac{\langle \hat{b}^\dagger \hat{b}^\dagger \hat{b} \hat{b} \rangle}{\langle \hat{b}^\dagger \hat{b} \rangle^2} = 2 + \frac{\delta\langle \hat{b}^\dagger \hat{b}^\dagger \hat{b} \hat{b} \rangle}{\langle \hat{b}^\dagger \hat{b} \rangle^2}. \qquad (A.10)$$

The time evolution of the correlation term is given by

$$(\hbar\frac{d}{dt}+4\kappa)\delta\langle\hat{b}^\dagger\hat{b}^\dagger\hat{b}\hat{b}\rangle = 4\left|g_{q_l,s}\right|^2\sum_{\nu'}\delta\langle\hat{b}^\dagger\hat{b}^\dagger\hat{b}\hat{v}^\dagger_{\nu'}\hat{c}_{\nu'}\rangle. \qquad (A.11)$$

The sum is again taken over all resonant laser transitions from various QDs. The photon correlation in term depends on the correlation between photon number and photon-assisted polarization $\delta\langle\hat{b}^\dagger\hat{b}^\dagger\hat{b}\hat{v}^\dagger_{\nu'}\hat{c}_{\nu'}\rangle$. The time-dependence of this term is already rather complicated:

$$\begin{aligned}(\hbar\frac{d}{dt}+3\kappa+\Gamma)\delta\langle\hat{b}^\dagger\hat{b}^\dagger\hat{b}\hat{v}^\dagger_{\nu'}\hat{c}_{\nu'}\rangle &= -2\left|g_{q_l,s}\right|^2\langle\hat{b}^\dagger\hat{v}^\dagger_\nu\hat{c}_\nu\rangle^2\\ &- (1-f^e_\nu-f^h_\nu)\delta\langle\hat{b}^\dagger\hat{b}^\dagger\hat{b}\hat{b}\rangle\\ &+ 2f^h_\nu\delta\langle\hat{b}^\dagger\hat{b}\hat{c}^\dagger_\nu\hat{c}_\nu\rangle - 2f^e_\nu\delta\langle\hat{b}^\dagger\hat{b}\hat{v}^\dagger_\nu\hat{v}_\nu\rangle\\ &- \sum_{\nu'}2\delta\langle\hat{b}^\dagger\hat{b}\hat{c}^\dagger_{\nu'}\hat{v}^\dagger_\nu\hat{c}_\nu\hat{v}_{\nu'}\rangle+\delta\langle\hat{b}^\dagger\hat{b}\hat{v}^\dagger_{\nu'}\hat{v}^\dagger_\nu\hat{c}_\nu\hat{c}_{\nu'}\rangle. \qquad (A.12)\end{aligned}$$

The last two terms can be neglected. The first of these is a generalization of the effect of Coulomb carrier correlations on the spontaneous emission source term which is negligible under usual conditions. The second of these requires annihilation of two conduction band electrons. As there is only two possible electron states with different spin values in the s-shell, this means that these electrons need to belong to different QDs if a circularly polarized cavity mode is considered. Correlations between carriers from different QDs correspond to superradiant coupling which can be neglected for the systems studied in this thesis. The remaining undefined terms are the photon-carrier correlations which already appeared in equation A.8. Their time-evolution is given by

$$\begin{aligned}(\hbar\frac{d}{dt}+2\kappa)\delta\langle\hat{b}^\dagger\hat{b}\hat{c}^\dagger_\nu\hat{c}_\nu\rangle &= -2\left|g_{q_l,s}\right|^2\Re[\delta\langle\hat{b}^\dagger\hat{b}^\dagger\hat{b}\hat{v}^\dagger_\nu\hat{c}_\nu\rangle\\ &+ \sum_{\nu'}\delta\langle\hat{b}^\dagger\hat{v}^\dagger_{\nu'}\hat{c}^\dagger_\nu\hat{c}_{\nu'}\hat{c}_\nu\rangle+(\langle\hat{b}^\dagger\hat{b}\rangle+f^e_\nu)\langle\hat{b}^\dagger\hat{v}^\dagger_\nu\hat{c}_\nu\rangle], \qquad (A.13)\\ (\hbar\frac{d}{dt}+2\kappa)\delta\langle\hat{b}^\dagger\hat{b}\hat{v}^\dagger_\nu\hat{v}_\nu\rangle &= 2\left|g_{q_l,s}\right|^2\Re[\delta\langle\hat{b}^\dagger\hat{b}^\dagger\hat{b}\hat{v}^\dagger_\nu\hat{c}_\nu\rangle\\ &- \sum_{\nu'}\delta\langle\hat{b}^\dagger\hat{c}^\dagger_{\nu'}\hat{v}^\dagger_\nu\hat{v}_{\nu'}\hat{v}_\nu\rangle+(\langle\hat{b}^\dagger\hat{b}\rangle+f^h_\nu)\langle\hat{b}^\dagger\hat{v}^\dagger_\nu\hat{c}_\nu\rangle]. \qquad (A.14)\end{aligned}$$

In these equations, the time evolution of all, but the sum terms has been given. However, these terms are of the same nature as the superradiant terms discussed before and can therefore be neglected.

The theoretical results for IO-curves and $g^{(2)}(0)$ of QD lasers shown in chapter 3 are numerical solutions of this set of equations. Calculations of results for $g^{(2)}(\tau)$ require more involved approaches. For these approaches, the quantities of interest like $\langle\hat{b}^\dagger(t)\hat{b}(t)\rangle$, $\langle\hat{b}^\dagger(t)\hat{b}^\dagger(t)\hat{b}(t)\hat{b}(t)\rangle$, and so on are first calculated in the steady state limit $t\to\infty$ according to the single-time equations of motion as discussed beforehand. It is now possible to calculate the τ-dependence of the unnormalized photon correlations:

$$G^{(2)}(\tau) = \langle\langle \hat{b}^\dagger(\tau)\hat{b}(\tau)\rangle\rangle \langle \hat{b}^\dagger \hat{b}\rangle \qquad (\text{A.15})$$

where the expectation value $\langle\langle\ldots\rangle\rangle$ is taken with respect to a modified density operator $\tilde{\rho} = \hat{b}\rho\hat{b}^\dagger/\langle \hat{b}^\dagger\hat{b}\rangle$. In this manner the two-time problem is reduced to two successively solvable single-time problems. Evaluation of the unnormalized photon correlations $G^2(\tau)$ requires using the single-time equations of motion for $\langle\langle \hat{b}^\dagger(\tau)\hat{b}(\tau)\rangle\rangle$, $\langle\langle \hat{c}_\nu^\dagger(\tau)\hat{c}_\nu(\tau)\rangle\rangle$ and the other quantities of interest with initial conditions

$$\langle\langle \hat{b}^\dagger(\tau)\hat{b}(\tau)\rangle\rangle|_{\tau=0} = \frac{\langle \hat{b}^\dagger\hat{b}^\dagger\hat{b}\hat{b}\rangle}{\langle \hat{b}^\dagger\hat{b}\rangle} \qquad (\text{A.16})$$

$$\langle\langle \hat{c}_\nu^\dagger(\tau)\hat{c}_\nu(\tau)\rangle\rangle|_{\tau=0} = \frac{\langle \hat{b}^\dagger\hat{b}\hat{c}_\nu^\dagger\hat{c}_\nu\rangle}{\langle \hat{b}^\dagger\hat{b}\rangle} \qquad (\text{A.17})$$

and so on. This approach assumes that the truncation introduced by the cluster expansion works equally well for operator averages taken with respect to the density operators ρ and $\tilde{\rho}$. The theoretical $g^{(2)}(\tau)$-curves shown in chapter 3 have been calculated using this approach.

Bibliography

[1] N. D. Mermin, American Journal of Physics **66**, 753 (1998).

[2] P. R. Rice, H. J. Carmichael, Phys. Rev. A **50**, 4318 (1994).

[3] T. Katsuyama, SEI Technical Review **69**, 13 (2009).

[4] V. Savona, L. C. Andreani, P. Schwendimann, A. Quattropani, Solid State Commun. **93**, 733 (1995).

[5] H. Deng, *Dynamic Condensation of Semiconductor Microcavity Polaritons*, Ph.D. thesis, Stanford university (2006).

[6] A. Nakamura, H. Yamada, T. Tokizaki, Phys. Rev. B **40**, 8585 (1989).

[7] K.D. Choquette, H.Q. Hou, Proceedings of the IEEE **85**, 1730 (1997).

[8] G. Khitrova, H. M. Gibbs, F. Jahnke, M. Kira, S. W. Koch, Rev. Mod. Phys. **71**, 1591 (1999).

[9] S. Pau, G. Björk, J. Jacobson, H. Cao, Y. Yamamoto, Phys. Rev. B **51**, 14437 (1995).

[10] M. S. Skolnick, T. A. Fisher, D. M. Whittaker, Semiconductor Science and Technology **13**, 645 (1998).

[11] G. Panzarini, L. C. Andreani, A. Armitage, D. Baxter, M. S. Skolnick, V. N. Astratov, J. S. Roberts, A.V. Kavokin, M. R. Vladimirova, M. A. Kaliteevski, Phys. Rev. B **59**, 5082 (1999).

[12] D. M. Whittaker, P. Kinsler, T. A. Fisher, M. S. Skolnick, A. Armitage, A. M. Afshar, M. D. Sturge, J. S. Roberts, Phys. Rev. Lett. **77**, 4792 (1996).

[13] T. Gutbrod, M. Bayer, A. Forchel, J. P. Reithmaier, T. L. Reinecke, S. Rudin, P. A. Knipp, Phys. Rev. B **57**, 9950 (1998).

[14] S. Reitzenstein, A. Forchel, Journal of Physics D: Applied Physics **43**, 033001 (2010).

[15] J.P. Reithmaier, G. Sek, A. Löffler, C. Hofmann, S. Kuhn, S. Reitzenstein, L. V. Keldysh, V. D. Kulakovskii, T. L. Reinecke, A. Forchel, Nature **432**, 197 (2004).

[16] A. Kavokin, J. J. Baumberg, G. Malpuech, F. Laussy, *Microcavities* (Oxford University Press, 2007).

[17] J. Kasprzak, S. Reitzenstein, E. A. Muljarov, C. Kistner, C. Schneider, M. Strauss, S. Höfling, A. Forchel, W. Langbein, Nature Materials **9**, 304 (2010).

[18] J. J. Hopfield, Phys. Rev. **112**, 1555 (1958).

[19] E. Giacobino, J.-P. Karr, A. Baas, G. Messin, M. Romanelli, A. Bramati, Solid State Communications **134**, 97 (2005).

[20] V. Savona, L. C. Andreani, P. Schwendimann, A. Quattropani, Solid State Communications **93**, 733 (1995).

[21] A. N. Kireev, M. A. Dupertuis, Optics Communications **123**, 268 (1996).

[22] Roy J. Glauber, Phys. Rev. **131**, 2766 (1963).

[23] U. Fano, American Journal of Physics **29**, 539 (1961).

[24] R. Hanbury Brown, R. Q. Twiss, Nature **178**, 1046 (1956).

[25] B. Chu, Ann. Rev. Phys. Chem. **21**, 145 (1970).

[26] C. Gutt, T. Ghaderi, V. Chamard, A. Madsen, T. Seydel, M. Tolan, M. Sprung, G. Grübel, S. K. Sinha, Phys. Rev. Lett. **91**, 076104 (2003).

[27] D. Huang, E. A. Swanson, C. P. Lin, J. S. Schuman, W. G. Stinson, W. Chang, M. R. Hoo, T. Flotte, K. Gregory, C. A. Puliafito, J. G. Fujimoto, Science **254**, 1178 (1991).

[28] Roy J. Glauber, Phys. Rev. **131**, 2766 (1963).

[29] R. Hanbury Brown, R. Q. Twiss, Nature **177**, 27 (1956).

[30] G. Li, T. C. Zhang, Y. Li, J. M. Wang, Phys. Rev. A **71**, 023807 (2005).

[31] D. J. Bradley, B. Liddy, W. E. Sleat, Optics Communications **2**, 391 (1971).

[32] L. M. Davis, C. Parigger, Measurement Science and Technology **3**, 85 (1992).

[33] S. M. Ulrich, C. Gies, S. Ates, J. Wiersig, S. Reitzenstein, C. Hofmann, A. Löffler, A. Forchel, F. Jahnke, P. Michler, Phys. Rev. Lett. **98**, 043906 (2007).

[34] S. Ates, S. M. Ulrich, P. Michler, S. Reitzenstein, A. Löffler, A. Forchel, Applied Physics Letters **90**, 161111 (2007).

[35] C. Gies, J. Wiersig, M. Lorke, F. Jahnke, Phys. Rev. A **75**, 013803 (2007).

[36] Perry R. Rice, H. J. Carmichael, Phys. Rev. A **50**, 4318 (1994).

[37] Y. Yamamoto, S. Machida, G. Björk, Phys. Rev. A **44**, 657 (1991).

[38] J. M. Gérard, B. Sermage, B. Gayral, B. Legrand, E. Costard, V. Thierry-Mieg, Phys. Rev. Lett. **81**, 1110 (1998).

[39] E. M. Purcell, H. C. Torrey, R. V. Pound, Phys. Rev. **69**, 37 (1946).

[40] M. Bayer, T. L. Reinecke, F. Weidner, A. Larionov, A. McDonald, A. Forchel, Phys. Rev. Lett. **86**, 3168 (2001).

[41] S. Reitzenstein, A. Bazhenov, A. Gorbunov, C. Hofmann, S. Münch, A. Löffler, M. Kamp, J. P. Reithmaier, V. D. Kulakovskii, A. Forchel, Applied Physics Letters **89**, 051107 (2006).

[42] J. Wiersig, C. Gies, F. Jahnke, M. Aßmann, T. Berstermann, M. Bayer, C. Kistner, S. Reitzenstein, C. Schneider, S. Höfling, A. Forchel, C. Kruse, J. Kalden, D. Hommel, Nature **460**, 245 (2009).

[43] M. Winger, T. Volz, G. Tarel, S. Portolan, A. Badolato, K. J. Hennessy, E. L. Hu, A. Beveratos, J. Finley, V. Savona, A. Imamoglu, Phys. Rev. Lett. **103**, 207403 (2009).

[44] S. Ates, S. M. Ulrich, A. Ulhaq, S. Reitzenstein, A. Löffler, S. Höfling, A. Forchel, P. Michler, Nature Photonics **3**, 724 (2009).

[45] M. Hennrich, *Kontrollierte Erzeugung Einzelner Photonen in Einem Optischen Resonator Hoher Finesse*, Ph.D. thesis, TU München (2003).

[46] M. Hennrich, A. Kuhn, G. Rempe, Phys. Rev. Lett. **94**, 053604 (2005).

[47] C. Gies, J. Wiersig, F. Jahnke, Phys. Rev. Lett. **101**, 067401 (2008).

[48] E. Jakeman, E. R. Pike, Journal of Physics A **2**, 115 (1968).

[49] H. Kurtze, J. Seebeck, P. Gartner, D. R. Yakovlev, D. Reuter, A. D. Wieck, M. Bayer, F. Jahnke, Phys. Rev. B **80**, 235319 (2009).

[50] S. I. Tsintzos, N. T. Pelekanos, G. Konstantinidis, Z. Hatzopoulos, P. G. Savvidis, Nature **453**, 372 (2008).

[51] D. Bajoni, E. Semenova, A. Lemaître, S. Bouchoule, E. Wertz, P. Senellart, J. Bloch, Phys. Rev. B **77**, 113303 (2008).

[52] F. Tassone, C. Piermarocchi, V. Savona, A. Quattropani, P. Schwendimann, Phys. Rev. B **56**, 7554 (1997).

[53] A. I. Tartakovskii, M. Emam-Ismail, R. M. Stevenson, M. S. Skolnick, V. N. Astratov, D. M. Whittaker, J. J. Baumberg, J. S. Roberts, Phys. Rev. B **62**, R2283 (2000).

[54] F. Tassone, Y. Yamamoto, Phys. Rev. B **59**, 10830 (1999).

[55] R. Houdré, J. L. Gibernon, P. Pellandini, R. P. Stanley, U. Oesterle, C. Weisbuch, J. O'Gorman, B. Roycroft, M. Ilegems, Phys. Rev. B **52**, 7810 (1995).

[56] J. R. Jensen, P. Borri, W. Langbein, J. M. Hvam, Applied Physics Letters **76**, 3262 (2000).

[57] V. Savona, F. Tassone, C. Piermarocchi, A. Quattropani, P. Schwendimann, Phys. Rev. B **53**, 13051 (1996).

[58] A. Imamoglu, R. J. Ram, S. Pau, Y. Yamamoto, Phys. Rev. A **53**, 4250 (1996).

[59] Le Si Dang, D. Heger, R. André, F. Bœuf, R. Romestain, Phys. Rev. Lett. **81**, 3920 (1998).

[60] S. Christopoulos, G. Baldassarri Höger von Högersthal, A. J. D. Grundy, P. G. Lagoudakis, A. V. Kavokin, J. J. Baumberg, G. Christmann, R. Butté, E. Feltin, J.-F. Carlin, N. Grandjean, Phys. Rev. Lett. **98**, 126405 (2007).

[61] R. Schmidt-Grund, B. Rheinländer, C. Czekalla, G. Benndorf, H. Hochmut, A. Rahm, M. Lorenz, M. Grundmann, Superlattices and Microstructures **41**, 360 (2007).

[62] J. Bloch, T. Freixanet, J. Y. Marzin, V. Thierry-Mieg, R. Planel, Applied Physics Letters **73**, 1694 (1998).

[63] A. J. Leggett, *Quantum Liquids - Bose Condensation and Cooper Pairing in Condensed-Matter Systems* (Oxford Press, New York, 2006).

[64] S. N. Bose, Z. Phys. **26**, 178 (1924).

[65] A. Einstein, Sitzungsber. Kgl. Preuss. Akad. Wiss. **261**, 3 (1924).

[66] O. Penrose, L. Onsager, Phys. Rev. **104**, 576 (1956).

[67] L. Onsager, Il Nuovo Cimento **6**, 279 (1949).

[68] R.P. Feynman, *Application of Quantum Mechanics to Liquid Helium*, vol. 1 of *Progress in Low Temperature Physics*, 17 – 53 (North Holland, Amsterdam, 1955).

[69] C. N. Yang, Rev. Mod. Phys. **34**, 694 (1962).

[70] P. W. Anderson, Rev. Mod. Phys. **38**, 298 (1966).

[71] J. Kasprzak, M. Richard, S. Kundermann, A. Baas, P. Jeambrun, J. M. J. Keeling, F. M. Marchetti, M. H. Szymanska, Andre. R., J. L. Staehli, V. Savona, P. B. Littlewood, B. Deveaud, Le Si Dang, Nature **443**, 409 (2006).

[72] H. Deng, G. S. Solomon, R. Hey, K. H. Ploog, Y. Yamamoto, Phys. Rev. Lett. **99**, 126403 (2007).

[73] S. Utsunomiya, L. Tian, G. Roumpos, C. W. Lai, N. Kumada, T. Fujisawa, M. Kuwata-Gonokami, A. Löffler, S. Höfling, A. Forchel, Y. Yamamoto, Nature Physics **4**, 700 (2008).

[74] K. G. Lagoudakis, M. Wouters, M. Richard, A. Baas, I. Carusotto, R. Andre, Le Si Dang, B. Deveaud-Pledran, Nature Physics **4**, 706 (2008).

[75] K. G. Lagoudakis, T. Ostatnicky, A. V. Kavokin, Y. G. Rubo, R. Andre, B. Deveaud-Pledran, Science **326**, 974 (2009).

[76] A. Amo, D. Sanvitto, F. P. Laussy, D. Ballarini, E. Del Valle, M. D. Martin, A. Lemaitre, J. Bloch, D. N. Krizhanovskii, M. S. Skolnick, C. Tejedor, L. Vina, Nature **457**, 291 (2009).

[77] A. Amo, J. Lefrere, S. Pigeon, C. Adrados, C. Ciuti, I. Carusotto, R. Houdre, E. Giacobino, A. Bramati, Nature Physics **5**, 805 (2009).

[78] J. J. Baumberg, A. V. Kavokin, S. Christopoulos, A. J. D. Grundy, R. Butté, G. Christmann, D. D. Solnyshkov, G. Malpuech, G. Baldassarri Höger von Högersthal, E. Feltin, J.-F. Carlin, N. Grandjean, Phys. Rev. Lett. **101**, 136409 (2008).

[79] H. Deng, G. Weihs, C. Santori, J. Bloch, Y. Yamamoto, Science **298**, 199 (2002).

[80] T. Horikiri, P. Schwendimann, A. Quattropani, S. Höfling, A. Forchel, Y. Yamamoto, Phys. Rev. B **81**, 033307 (2010).

[81] J. Kasprzak, M. Richard, A. Baas, B. Deveaud, R. André, J.-Ph. Poizat, Le Si Dang, Phys. Rev. Lett. **100**, 067402 (2008).

[82] P. C. Hohenberg, Phys. Rev. **158**, 383 (1967).

[83] N. D. Mermin, H. Wagner, Phys. Rev. Lett. **17**, 1133 (1966).

[84] S. Coleman, Communications in Mathematical Physics **31**, 259 (1973).

[85] J. M. Kosterlitz, D. J. Thouless, Journal of Physics C: Solid State Physics **6**, 1181 (1973).

[86] A. P. D. Love, D. N. Krizhanovskii, D. M. Whittaker, R. Bouchekioua, D. Sanvitto, S. Al Rizeiqi, R. Bradley, M. S. Skolnick, P. R. Eastham, R. André, Le Si Dang, Phys. Rev. Lett. **101**, 067404 (2008).

[87] T. D. Doan, H. T. Cao, D. B. Tran Thoai, H. Haug, Phys. Rev. B **78**, 205306 (2008).

[88] P. Schwendimann, A. Quattropani, Phys. Rev. B **77**, 085317 (2008).

[89] D. Sarchi, P. Schwendimann, A. Quattropani, Phys. Rev. B **78**, 073404 (2008).

[90] H. Deng, D. Press, S. Götzinger, G. S. Solomon, R. Hey, K. H. Ploog, Y. Yamamoto, Phys. Rev. Lett. **97**, 146402 (2006).

[91] G. Roumpos, C.-W. Lai, T. C. H. Liew, Y. G. Rubo, A. V. Kavokin, Y. Yamamoto, Phys. Rev. B **79**, 195310 (2009).

[92] M. Wouters, I. Carusotto, C. Ciuti, Phys. Rev. B **77**, 115340 (2008).

[93] M. Wouters, I. Carusotto, Phys. Rev. Lett. **99**, 140402 (2007).

[94] M. H. Szymańska, J. Keeling, P. B. Littlewood, Phys. Rev. Lett. **96**, 230602 (2006).

[95] I. Carusotto, C. Ciuti, Phys. Rev. Lett. **93**, 166401 (2004).

[96] I. Carusotto, C. Ciuti, Phys. Rev. B **72**, 125335 (2005).

[97] D. Porras, C. Ciuti, J. J. Baumberg, C. Tejedor, Phys. Rev. B **66**, 085304 (2002).

[98] D. Ballarini, D. Sanvitto, A. Amo, L. Viña, M. Wouters, I. Carusotto, A. Lemaitre, J. Bloch, Phys. Rev. Lett. **102**, 056402 (2009).

[99] M. Richard, *Quasi-Condensation of Polaritons in II-VI CdTe-Based Microcavities under Non-Coherent Excitation*, Ph.D. thesis, Université Joseph Fourier (2004).

[100] C. Ciuti, V. Savona, C. Piermarocchi, A. Quattropani, P. Schwendimann, Phys. Rev. B **58**, 7926 (1998).

[101] M. Combescot, O. Betbeder-Matibet, R. Combescot, Phys. Rev. B **75**, 174305 (2007).

[102] Y. Shinozuka, M. Matsuura, Phys. Rev. B **28**, 4878 (1983).

[103] Y. G. Rubo, A.V. Kavokin, I.A. Shelykh, Physics Letters A **358**, 227 (2006).

[104] N. Baer, C. Gies, J. Wiersig, F. Jahnke, European Physical Journal B **50**, 411 (2006).

[105] J. Fricke, Annals of Physics **252**, 479 (1996).

[106] M. Kira, F. Jahnke, W. Hoyer, S. W. Koch, Progress in Quantum Electronics **23**, 189 (1999).

[107] W. Hoyer, M. Kira, S. W. Koch, Phys. Rev. B **67**, 155113 (2003).

[108] M. Schwab, H. Kurtze, T. Auer, T. Berstermann, M. Bayer, J. Wiersig, N. Baer, C. Gies, F. Jahnke, J. P. Reithmaier, A. Forchel, M. Benyoucef, P. Michler, Phys. Rev. B **74**, 045323 (2006).

[109] T. Feldtmann, L. Schneebeli, M. Kira, S. W. Koch, Phys. Rev. B **73**, 155319 (2006).

[110] T. R. Nielsen, P. Gartner, F. Jahnke, Phys. Rev. B **69**, 235314 (2004).

List of Figures

1.1 Semiconductor electronic density of states under varying dimensionality. 10
1.2 Transition from weak to strong coupling. 11
1.3 Rabi oscillations. 16
1.4 Eigenmodes of a system of two linear coupled oscillators. 18
1.5 Hopfield coefficients. 19
1.6 Angle-dependent Hopfield coefficients and polariton lifetimes. 21
1.7 Detuning-dependent polariton dispersion. 22
1.8 Splittings seen in absorption, transmission, reflectivity and PL. 23
1.9 $g^{(2)}(\tau)$ for thermal, coherent and nonclassical light. 25
1.10 Photon bunching explained by two-photon interference. 29
1.11 Photon number distributions for coherent, thermal and Fock states. 30

2.1 Schematic TCSPC measurement. 35
2.2 Overview of the experimental setup. 37
2.3 Close-up view of the laser system. 38
2.4 Close-up view of the polarization optics. 39
2.5 Close-up view of the microscope objective. 40
2.6 Close-up view of the cryostat. 41
2.7 Close-up view of the optics in the detection path. 42
2.8 Close-up view of the monochromator. 44
2.9 Close-up view of the streak camera. 45
2.10 Single and integrated streak camera images. 48
2.11 Influence of the detector temporal resolution on the measured $g^{(2)}$. 50
2.12 Influence of detector noise on the measured $g^{(2)}(0)$. 51
2.13 Jitter-broadenend photon pair count rates. 54
2.14 Jitter-induced intensity autocorrelation trace of a laser pulse. 55
2.15 Jitter dependence of the equal-time correlation functions. 57

2.16 Jitter dependence of $g_j^{(2)}(0,0)$. 59

2.17 Effect of jitter on $g_j^{(2)}(t,0)$. 60

3.1 Input-output curves for lasers with increasing β-factor. 62

3.2 Mode spectrum of the low-Q micropillar laser. 64

3.3 IO curves and $g^{(2)}(0)$ for three different QD lasers. 65

3.4 Calculated $g^{(2)}(0)$ and IO-curve for a low-Q micropillar laser. 67

3.5 Calculated $g^{(2)}(0)$ and IO-curve for a high-Q micropillar laser. 68

3.6 Temporal evolution of $g^{(2)}$ for selected pump powers. 69

3.7 Calculated $g^{(2)}(\tau)$ for a low-Q micropillar laser. 70

3.8 Calculated $g^{(2)}(\tau)$ for a high-Q micropillar laser. 71

3.9 Integrated intensity of the 6 μm pillars fundamental mode. 73

3.10 Measured time evolution of $g^{(2)}(t,0)$ for a low-Q micropillar. 74

3.11 Relative fractions of coherent and thermal emission at fixed $g^{(2)}(t,0)$. 76

3.12 Effects of common and rare jitter events on $g^{(2)}(t,0)$. 77

3.13 Calculated IO-curve and $g^{(2)}(t,0)$ for a low-Q micropillar. 79

4.1 Spontaneous polariton relaxation mechanisms. 82

4.2 Dispersions at the QW-diode to VCSEL transition. 84

4.3 IO-curve and emission blueshift at the QW-diode to VCSEL transition. 85

4.4 Higher-order correlation functions at the QW-diode to VCSEL transition. 86

5.1 Measured $g^{(2)}(0)$ and $g^{(3)}(0)$ of a polariton BEC ground state. 96

5.2 Measured $g^{(2)}(0)$ of a single-polarization polariton BEC ground state. 98

5.3 Polariton dispersions for varying detunings and excitation densities. 100

5.4 Flat polariton dispersion under nonresonant pumping. 102

5.5 Calculated condensate properties assuming a parabolic dispersion. 106

5.6 Calculated condensate properties assuming a Bogoliubov dispersion. 107

5.7 Momentum-space intensity profile as predicted by different models. 110

5.8 Comparison of quantum and thermal depletion. 112

5.9 Quantum and thermal depletion considering higher-order contributions. 115

Index

$g^{(2)}$, 24

Bogoliubov quasiparticle, 111
Bogoliubov transformation, 17
Bose-Einstein condensation, 87

CCD, 50
cluster expansion method, 120

DBR, 13
diffusive Goldstone mode, 102, 114
dressed states, 16

Feynman-Onsager condition, 91

Gross-Pitaevskii equation, 103

Hanbury-Brown-Twiss setup, 34
highly photonic states, 95
Hopfield coefficients, 19

local density approximation, 104

off-diagonal long-range order, 92

Pauli blocking, 121
photon-assisted polarization, 121
polariton OPO, 105
Purcell factor, 63

quantized vortex, 92
quantum depletion, 111

Rabi oscillations, 16
Rabi splitting, 10, 22

relaxation bottleneck, 81
relaxation-time approximation, 121
semiconductor luminescence equations, 120
single photon avalanche photodiode, 34
Spontaneous symmetry breaking, 92
strong coupling, 10, 15
superfluid velocity, 91

thermal depletion, 94
thermodynamic limit, 92
time-correlated single photon counting, 33

VCSEL, 61

weak coupling, 10, 24

Symbols and abbreviations

symbol	meaning
a_B	2D exciton Bohr radius
β	spontaneous emission factor
BEC	Bose-Einstein condensate
c	speed of light ($299792458\,\mathrm{m/s}$)
CCD	charge-coupled device
CW	continuous wave
DOS	density of states
e	electron, unit charge ($1.602176 \cdot 10^{-19}$ C)
E_b	exciton binding energy
E_{gap}	band gap energy
eV	electron volt ($1.602176 \cdot 10^{-19}$ J)
f	focal length
F_P	Purcell factor
FWHM	full width at half maximum
$g^{(2)}$	second order correlation function
$g^{(3)}$	third order correlation function
GPE	Gross-Pitaevskii equation

$\hbar$	$h/2\pi$ = 1.054571· 10^{-34} J s= 6.582118· 10^{-16} eV s
HBT	Hanbury-Brown-Twiss
k_B	Boltzmann constant ($1.38062 \cdot 10^{-23} JK^{-1}$)
$\vec{k}$	wave vector
$k_{\|}$	in-plane wave vector
K	Kelvin
λ	wavelength
L_{eff}	effective cavity length
LP	lower polariton
m_0	free electron mass ($9.109381 \cdot 10^{-31}$ kg)
MCP	micro-channel plate
meV	milli-electron volt
μm	micrometer
M^{X-X}	exciton-exciton interaction matrix element
n	refractive index
NA	numerical aperture
Nd:YAG	yttrium aluminium garnet doped with neodymium
nm	nanometer
ODLRO	off-diagonal long-range order
PL	photoluminescence
ps	picosecond
ψ	order parameter
Q	quality factor
QD	quantum dot
QW	quantum well

R	reflectivity
τ_c	coherence time
t	time
T	temperature
TCSPC	time-correlated single photon counting
UP	upper polariton
VCSEL	vertical-cavity surface-emitting laser
$\vec{v}_s$	superfluid velocity
W	Watt

Acknowledgements

Finally, it is time to thank all the people without whom this thesis would not have been possible at all.

I would like to thank Prof. Dr. Manfred Bayer for giving me the opportunity to perform all these studies in one of the best-equipped labs there is and for the guidance and invaluable help.
I am grateful to Priv.-Doz. Dr. Dmitri R. Yakovlev and Prof. Dr. Dietmar Fröhlich for sharing some of their experience on many devices, experimental setups and spectroscopy in general.
Thanks to Thorsten Berstermann, Jean-Sebastian Tempel, Franziska Veit and Lars Erik Kreilkamp for spending some time with me on the streak camera setup in some of the four labs the setup has been located at in the last years. Looking for that special micropillar and sneaking out of and back into the lab hoping the setup keps stable is much less frustrating when you are not alone.
Further I would like to thank all the people we collaborated with in the recent few years. Vicariously for all the collaborators let me name Dr. Stephan Reitzenstein and Prof. Dr. Forchel from the Würzburg epitaxy and spectroscopy group, Prof. Dr. Hommel from the Bremen epitaxy group, Prof Dr. J Hvam from the DTU Fotonik and Dr. Christopher Gies and Prof. Dr. Jahke from the Bremen theory group.
I am grateful to Klaus Wiegers, Thomas Stöhr and Michaela Wäscher. I am quite convinced they solved more technical and administrative problems for me than I noticed and surely more than I asked them to.
Further I would like to thank all members of Experimentelle Physik 2 - meanwhile too many to name everybody here - for the familiary atmosphere and lots of fun.
Thanks to all my friends for reminding me of the existence of a world outside of the university. Special thanks to the guys from the Y. Y thank you all for beyng the mayn reason why there

ys no wednesday.

I owe a lot to my family. I thank my mother, Christine Aßmann and my sister Sabine Grieser and my brother-in-law Sascha Grieser for their support and care.

Finally I would like to thank Monika Lippert for bearing with me and not questioning that strange stuff I spent lots of time on.

Die VDM Verlagsservicegesellschaft sucht für wissenschaftliche Verlage abgeschlossene und herausragende

Dissertationen, Habilitationen, Diplomarbeiten, Master Theses, Magisterarbeiten usw.

für die kostenlose Publikation als Fachbuch.

Sie verfügen über eine Arbeit, die hohen inhaltlichen und formalen Ansprüchen genügt, und haben Interesse an einer honorarvergüteten Publikation?

Dann senden Sie bitte erste Informationen über sich und Ihre Arbeit per Email an *info@vdm-vsg.de*.

Sie erhalten kurzfristig unser Feedback!

VDM Verlagsservicegesellschaft mbH
Dudweiler Landstr. 99   Telefon +49 681 3720 174
D - 66123 Saarbrücken   Fax   +49 681 3720 1749

www.vdm-vsg.de

Die VDM Verlagsservicegesellschaft mbH vertritt

Printed by Books on Demand GmbH, Norderstedt / Germany